Anonymus

The banditti of the Rocky mountains and Vigilance committee in

Idaho

Anonymus

The banditti of the Rocky mountains and Vigilance committee in Idaho

ISBN/EAN: 9783742831477

Manufactured in Europe, USA, Canada, Australia, Japa

Cover: Foto ©Klaus-Uwe Gerhardt /pixelio.de

Manufactured and distributed by brebook publishing software
(www.brebook.com)

.

Anonymus

The banditti of the Rocky mountains and Vigilance committee in Idaho

The
anditti of the Rocky Mountains
and
Vigilance Committee in Idaho

An Authentic Record
of Startling Adventures
In the Gold Mines of Idaho

Anonymous

gang, as the good citizens have been able to bring to justice.

With these premises, we ask the reader to turn to the subject and wander for awhile with us through the beautiful valleys and mountain gorges of Idaho, hoping in our rambles to both interest and satisfy all.

Contents

Banditti of the Rocky Mountains

OR

THE MURDERERS DOOM

A TALE OF THE GOLD MINES

CHAPTER I

The Rocky Mountains, from their earliest settlement, have probably been more widely noted as being infested with a class of reckless, blood-thirsty ruffians, than any other section of our whole country, if indeed it does not vainly challenge the early settlement of the old world for a parallel—owing, perhaps, to two great causes: the first of which is, its being so remote from civilization, and the second is, its being peculiarly adapted to their hellish trade.

Located as this wonderful chain of mountains is, in the interior of the continent, with the great American Desert stretching from its either base hundreds of miles, rendered the pursuit and detection of those, who have the hardihood to venture among its hostile tribes and sandy wastes, almost impossible.

Thus has nature, as it were, secured the safe transit of countless desperadoes and law-doomed men, to more congenial localities, amid the mountain gorges and passes, where the ever alluring gold attracts the sturdy miner, followed by merchants, speculators and business men of every description, who not unfrequently fall unsuspecting victims into the hands of these incarnate devils, whose business it is to gorge their own pockets with plunder purchased with the blood of their fellow man; and thus it is that many a mountain gorge and canon bears the name of some murdered man, who fell a victim to avarice, and whose resting place can never be known until the last great trump shall "quicken both the living and the dead."

Bands of desperadoes were not only organized in California, in its earliest and most palmy days, but also in Nevada, Utah and Colorado, that have rode rough-shod, as it were, over the rights of the people, trampling upon both human and divine law, and so disturbing the quiet of the peaceful citizens, that not unfrequently Judge Lynch has adjudicated, and a signpost or a neighboring tree, has been converted for the time being, into a gallows, before which the doomed victim has plead unavailingly for mercy.

However much may be said in favor of thus speedily ending the mad career of those who prey upon the innocent, one thing is certain, and that is this: that while it generally has a salutary effect for a time upon that class of men, in the way of rendering them less bold, or perhaps, if you please, more discreet, it usually disconcerts their plans, and renders new fields of operation indispensable to their success. Yet, it seldom, if ever, reforms any who escape its condign punishment, as they soon congregate in more remote localities, where new theaters of operation await the outburst of their pent up malice, stung by recent

2

defeats and persecutions, and a new community is startled with the awful cry of murder in their midst.

The object of this compilation is not to give a full account of all the numerous murders and robberies that have been perpetrated in Idaho, by banditti infesting that territory, as that would be too great a task, and require more space than we could give in this little volume. Hence, the reader will be content with perusing such facts as are connected with some of the most shocking murders committed by this lawless band, with an authentic biography of its captain, (Plummer) and other notorious characters connected with the gang.

We trust our readers will not be so terrified upon being introduced into the drinking saloons and gambling hells, which must form no insignificant portion of the tale of blood and woe, which it is the object of the writer to reveal to the public, as to turn in disgust from these sinks of crime, but that they will go with us for the time being, into those chambers of depravity and death, and there review the character and bloody career of such men as the notorious Plummer, who became captain of the banditti of the mountains.

To visit him in his most secret haunts; to expose the crime to which he was either principal or accessory; to lay before the world his private, as well as secret and public character; and to trace the blood-stained record of Slaid , whose notoriety in the world of crime is as wide as the continent, and whose character is as dark as the region to which he has gone. To trace these two characters and some of their leading accomplices, fearlessly, through the bloody

3

tragedies of California and Idaho, will not only require the full particulars connected with the murder of Venard, of California, and George Coply and Lawrence Kealey, of Bannock City, Magruder, of Lewiston, and the Deputy Sheriff of Virginia City; but also the full particulars of the trial and execution of Sheriff Plummer, Slaid, Buck Stinson, Ned Ray, Jack Gallager, Boon Helm, Dutch John, a Spaniard, and some thirty others of the gang.

Thus having stated our object in presenting to the public this little volume, and the character of the tale which we purpose to relate, we enter unhesitatingly upon the task, believing we shall be sustained and amply repaid for all the time and money we have expended in the arduous task of procuring the data and facts herein recorded.

In the autumn of 1862, gold was discovered upon the head waters of the Missouri River, by a party whom Captain Fisk , of the United States Army, had escorted across the country from St. Paul that season. The country for hundreds of miles in extent, in every direction, from the mines, was unsettled and uninhabited, save by the hostile Indian and the Buffalo. The discoveries soon became known throughout the neighboring territories and states, where tales of golden wealth were whispered from ear to ear, and imaginary fortunes accumulated, as it were, by the eager multitude that emigrated thither, ere winter had set in, and who were so soon to feel the horrors of terror from the presence of lawless ruffians and "scape gallows," who had also been allured thither from their former base of operations, hoping to be able, in the absence of law, either civil or military,

4

to carry on their nefarious and hellish traffic under more favorable auspices.

The little city of Bannock sprung, magic like, into existence, situated upon a tributary of the Missouri, in a beautiful valley at the head of a mountain gorge. The stroke of the hammer, the clatter of hoofs, and the steady tread of the miner, but told too well that a change had come over the awful silence of this mountain valley, where naught save the zephyrs that whistle through the nodding pine upon the mountain side, or the rippling waters of the brooklets, as they glide over their pebbly bottoms, had broken in upon its death-like stillness, since the great up-heavings of chaos. Now all was bustle, all was life. Men of every clime and nation were here. Cabins, stores and hotels were the growth of a night. Saloons, billiard tables and monte banks occupied the business corners.

The mines soon began to yield up their golden treasures, and business men assumed an air of well-to-do, and thus matters continued all moving quietly on for some little time. It was soon discovered however, that the town was infested with a gang of roughs, and the exchange of pistol shots soon became common occurrence. However, little blood was shed during the following winter. On one occasion a little shooting affray took place in Skinner's saloon, in which one man was killed, also one man and one boy wounded. The circumstances are as follows:

Four desperadoes were gambling at a euchre table, the stakes being fifty dollars on a side. The charge of cheating was made to one Reeves, who denied it with a threat, accompanied with an oath, whereupon one of his opponents seized the stakes. This brought

Reeves to his feet, with pistol in hand, which was immediately leveled and fired at his antagonist. The firing now becoming general, the bystanders hastily betook themselves to the door for safe keeping, but the infuriated gamblers continued to fire away at each other, until they had unloaded their revolvers, and strange to say, all the combatants escaped unharmed —owing, perhaps, to their being intoxicated—save one, who was wounded in the leg. A boy of about the age of fourteen, who was passing the door at the time of the shooting, was also wounded in the leg, and crippled for life, while a stranger, who was laying at the time upon the counter of the bar, in a state of intoxication, was also shot, the ball passing through the chest and killed, dying instantly, and without a struggle.

Thus, for the first time, blood had been shed in Bannock, and the horrible tale of murder was whispered in death-like silence from cabin to cabin, among the more peaceful and timid. The death-knoll had been sounded, and the first victim surrendered to the guillotine of vice. Business men now walked the streets with a quicker tread, and the miners remained in their cabins after night fall, while women were seldom seen upon the streets. All apparently conscious of the danger to which they were daily exposed, from the presence of these heartless ruffians.

An effort was soon made, however, to bring the guilty parties to justice, which failed, almost in its inception, from the fact that, as yet, no civil organizations had been completed, and consequently there was no one to take the lead, and the result was, such as might have been expected, namely, the parties being simply banished from the country, by a vote of the miners, which amounted to nothing, as some of them returned the third day after leaving the city.

For some time subsequent to this exhibition of despotism, little of importance occurred in Bannock in the way of blood letting. Men who had been in the habit of spending their evenings at the monte bank or around the euchre table , in these gambling hells, where the spots upon an upturned dice, or the drawing of a card, determined the fortunes of the hour, now sought retirement in the more peaceful avocations of life. But there was another class of men for whom murder in its darkest form has no terror. This class now ruled the hour at Bannock, and almost every stage coach that arrived, brought some miserable renegade from the states or territories, and thus accessions to their ranks were continually being made.

The winter dragged its weary length along with no change for the better. Nights of debauchery and gambling, with now and then the exchange of a few shots or the killing of a man, made up the programme of this gang for the winter.

But spring came, and with it a disposition upon the part of the good citizens to establish civil law, for the purpose of protecting themselves against lawless depredations, and to bring to justice those who recognized neither the laws of God or man.

Here we must beg leave of our readers to depart from the present subject, and take a detour with the notorious Plummer , who subsequently became sheriff of Bannock City, and also captain of the banditti of Idaho, from the peaceful home of his youth in Wisconsin, to the broad and extensive theater of blood, upon which he acted so conspicuous a part in California. After which, we shall show the reader how sadly the good people of Bannock were disappointed in the officers elected to administer civil law.

CHAPTER II

In the rural district in the state of Wisconsin, is situated a beautiful cottage, in which resided the parents of young Plummer.

It was a lovely morning in the month of May in the spring of 1852. The sun had risen, and was now pouring its gilding rays through the time-worn window, into the room occupied by the youth whose name was subsequently to become a terror throughout the Pacific slope, not only to the defenceless, but also to men of great daring and bravery.

In his room, he sat quietly adjusting many little articles, now objected to some as being to cumbersome, now accepting others as they were presented, as the last offering of a kind mother, and snugly packing them away in his valise, preparatory to making his final exit from beneath his father's roof. Long had it shielded him from the cold chilly world, without, with its blessings and its protection. Long had the silvery whisperings of maternal affection fell

9

upon his youthful ear, and the blessings of a kind father been lavished upon his devoted head.

Whatever of means it had required to educate him suitable to his condition in life, had been freely given, and it had not been as "casting pearls before swine," for the youth had appreciated these early privileges, and now stood before the world both a refined gentleman and a scholar, beautified with all the graces that adorn his sex under most favorable circumstances.

Golden tales now allure him to the Pacific coast. His parents kindly remonstrate, but in vain. Brothers, sisters, and old familiar friends gather around, and vainly implored that he shall not break the sunny circle of youth. But his determination is fixed, and all the tears and entreaties from whatever source, seem to fall as dead weight before an immovable heart and a determined will.

The time for his departure drew nigh. The youth resolutely bade farewell to all he held dear, and seated himself in the coach, and as it bore him slowly but surely away, he arose and gazingly took one long lingering look at the old homestead, and sank back upon his seat, fully conscious, for the first time in life, that he was sailing upon its stormy sea, without a home to shield him from its angry waves, or a parent to council him in his wanderings.

The next trace we have of him, is among the auriferous hills and golden valleys of California. He is now trying the stern realities of life, in a foreign clime, among strangers. The first season he passes in fruitless labor in Sacramento Valley, beneath the scorching rays of a Pacific sun. His golden hopes have not been realized, and disappointment settles

10

upon his youthful brow. Unfamiliar faces meet his gaze, and other associates than those of youth gather around, and he is urged to spend the coming winter in San Francisco. Here he took the first lesson in the school of vice, that was to educate him for the gallows.

Disappointed in mining operations, he sought employment in the city. The meager salary of seventy-five dollars per month, as a clerk, did not long satisfy his insatiate desire for money. Meantime he had observed how readily fortunes were made and lost by the turning of a card, in this great gold mart of the continent. He, too, had occasionally dropped into some of the more fashionable gambling houses, and invested a few dollars at a monte bank, roulette, or euchre table, with signal success, and now that he had made a raise, why not enter into some business that would be less laborious and more lucrative than clerking.

Thus he argued to himself, as he stood upon the street corner. There is Mr. A, said he,—a friend of mine—who made ten thousand at the euchre table last night, and Mr. B, who made five thousand at a monte bank, in the short space of an hour, while I was selling a few musty goods to one of our customers this morning.

Thus he run over the names of several whom he knew to have made money at gambling within the last few days. He had left his early home and the peaceful avocations of civil life, for the sole purpose of attaining wealth, and now with the propheticness of youth, he saw the goal of his ambition attained most speedily through a life at the gambling table.

Yes, said he, with an air of seriousness, for he was a man whose natural impulses were of no ordinary character. Yes! he unhesitatingly repeated, I will call for a settlement with my employer to-night. Thus upon the gilded rock of temptation his bark floundered, and not a deed or act of blood in all his after life, but may be traced back to this very spot.

The settlement was made, and the following night Plummer might have been seen at the Bank Exchange, "taking chances," as he termed it, at a gambling table. We will not attempt to follow him through all the varied meanderings of vice and fortune of the two succeeding years, but hastily pass on and take a glance at some of the most important events of his life.

In the year '55, we find him in Nevada City Up to this time, on more than one occasion, his hands had been imbrued in human gore, and long since he has borne the name of a desperado. At this time, Nevada City was a mere town just springing into existence, situated in a rich mining district, in Nevada county, California. Like all other rural towns of those early days in California, it was infested with gamblers, robbers and murderers, many of whom had escaped the wrath of Vigilance Committees in other localities.

An election was soon to take place in the city, and Plummer being the acknowledged bravado for reckless daring, was placed upon the "bummer ticket" for City Marshal, and what his party lacked in votes, they made up on election day, by threats and lawless assaults upon respectable citizens, thus preventing a full vote being taken, and giving the election to Plummer. At the expiration of his first term of office,

12

he was re-elected, and subsequently run for the California legislature, upon the democratic ticket, while he remained in the city, but was defeated.

It may justly be said of him, that as a public officer, he was not only prompt and energetic, but also kind to such as the duties of office compelled him to oppress, though when opposed in the performance of his official duties, he became as bold and determined as a lion, carrying out the orders and decrees of the court with the fierceness of a tiger, yet scrupulously exact in all business transactions, and performing the duties of his office with signal ability.

Business integrity had gained him many friends, even among the more respectable classes of society. And while his official garb had enabled him to more effectually carry on his hellish trade, it had also introduced him—in a degree—into the higher circles of society, where many regarded him with an eye of pity, rather than otherwise, considering him a fallen monument of early rectitude and genius. During his administration as marshal, he formed the acquaintance of a very amiable and beautiful young lady, the daughter of one Mr. B., an extensive merchant of the city, to whom, after months of association, he became engaged, though against the expressed will of her parents, and but for subsequent developments would doubtless have made her his wife.

The turbid waters of iniquity, however, could not long continue to sweep on in their mad career, under an official cloak. It has wisely been said, that "there is a change in the tide of every man's life." That change with Plummer had now come, and the city was aroused from its midnight slumbers, by the echo of pistol shots, followed by the sickening shriek of murder.

Plummer had killed an innocent and defenceless man in his own house. During that season he had formed the acquaintance of a Mrs. Venard , who resided in the city, and whose character had always been considered spotless, until seduced by the cunning arts and wiles of Plummer, into an unsuspected and unauthorized intimacy, which not only created sus- picions in the minds of her neighbors, as to her virtuous character, but also aroused the passions of jealousy in the breast of her husband.

The intimicy between the parties, continued for some months to become more notorious, Plummer frequently taking her out riding, and also to parties, during the absence of her husband, who was doing business in the mines, some distance from the city, and did not come home only once or twice a week, but on the fatal night he had come at an unusual late hour. Entering his house quietly and unsuspectingly, he spoke to his wife, calling her by her maiden name, (Martha) but no voice responded. Supposing her to be asleep, he stepped across the room and opened her bed-room door, whereupon Plummer leaped from the bed, seizing his revolver, and commenced firing at Mr. Venard, who, with a cry of murder, sought to make his escape, but was pierced with four balls, and fell a lifeless corpse upon his own door-steps. Thus the very fiend who had undermined and robbed him of his domestic happiness, had also brutally assassinated him in his own temple, in the presence of his once loved Martha.

The cry of murder and the echo of pistol shots, at

14

Arrest of Plummer.

the dead of night, in the suburbs of the city, away from the gambling and drinking hells—and what was more strange, that it should be at the house of Mr. Venard, a gentleman of reputed sobriety and quietness—aroused those of the immediate neighborhood, and soon a large crowd of people were assembled at the house. Mr. Venard was found lying upon his door-steps, pierced with bullets, and weltering in his own gore, which was still gushing warmly from his heart.

Plummer was seized by a policeman, while endeavoring to make his escape through the back window of Mrs. Venard's sleeping room, and dragged off to a police station, very much against the wishes of the incensed crowd, who sought, on more than one occasion, while on the way to the station, to wrest him from the police force, which had now become quite strong, for the purpose of summarily suspending him to a lamp-post. The crowd was, however, persuaded to desist, from the assurances given by the police, that justice should be done in the case. A large throng continued to hang around the station during the night, and it was with the greatest difficulty—as the crowd augmented—that they were prevented from breaking through and dealing swift and terrible retribution to the deserving fiend within.

For this heinous and almost unparalleled crime, Plummer was formally tried, but not convicted, from the fact that it was proven, in the course of the trial, that Mr. Vanard had, while in conversation with certain parties, with regard to the intimacy of his wife and Plummer, publicly said that he would shoot Plummer, on sight. This being considered by the jury

16

a reasonable excuse for the commission of the horrible crime, notwithstanding Plummer was the first aggressor, and might have been killed by Venard, without incurring the penalty of the law, either human or divine.

The day subsequent to the development of this crime, Plummer received the following note, while in jail, from Miss B——, the lady of whom we have before spoken.

Nevada City, October 15, 1858.

Once Loved Henry:——

The dreadful news of your crime reached me a few hours since. Need I say, at first it fell like a bolt upon my heart. Since no language can describe my present feelings.

The golden chain of love that has entwined our hearts for weeks that have passed, is forever severed. No longer can I think of you as a lover. Your crime, the foulest of foul, is not to be forgiven by the God who rules above, much less by such as are of earth. I am the deceived, and alone must regret our former acquaintance—you that should have acted a nobler part—the deceiver. Too well have you acted your base part—that of a lover. Perhaps you thought to still keep your crimes from the gaze of the world, and continue your plan of deception to one you thought could see no guile—if see, would forgive. But base one, your crime is not a secret, and though I have loved you, I love you no more, and pray the vengeance of God upon you. May you receive a murderer's reward, not only in this world, but that beyond. Methinks a murderer's death far too honorable for you to die. You may start at such a note from me, but you

17

need not. My love has turned to hate, and too well I know now that you never was worthy of the love bestowed upon you, but more worthy such a being to dwell in the regions of Pluto, than among the fair of this earth. I have no more to say, but leave you in the hands of justice.

Anna.

Notwithstanding this severe rebuke from his intended, and the indignation that was now every where exhibited towards him, he still continued to remain in town, but appearing more indifferent however, as to the publicity of his daily vice and crime.

Now that Miss B. had disregarded him, he apparently lost all self-respect, and betook himself to the low haunts of iniquity, (houses of ill-fame) where he passed much of his time. He soon formed an attachment to a young woman, who resided in one of those dens of infamy, that was kept by one Ashman , and for some months waited upon her at balls and other places of amusement, she going by the name of Madame Plummer.

To say that she was a person of exquisite beauty, would not express half that should be said with regard to her angel-like appearance. A form, tall and graceful, well proportioned in all its outlines, with exquisitely formed features, while golden ringlets of auburn hung upon her shoulders, and the piercing glances of her dark eyes bespoke for her something more than ordinary endowments. Possessed of such charming beauty, it would be idle to say that she was without admirers, although an inmate of one of those houses to which so many young and promising females in an evil hour have been seduced.

Among her admirers, and the most prominent rival of Plummer, was one Riley , who had been fascinated by her winning smiles, and charmed by her graces, until an intimacy had grown up between the two, which was any thing but pleasant to Plummer, who vainly sought to persuade her to discard Riley, which she declined doing, but on the contrary accepted him as her master, doubtless because of his wealth, she having already exhausted the resources of Plummer.

This aroused the passions of jealousy in her former master, who, upon meeting his rival at Ashman's house, soon got into difficulty with him, when a severe altercation took place, resulting in the death of Riley, which was occasioned by a pistol shot fired by Plummer. For this offence he was tried, condemned and sentenced to be hung, but during the night previous to the day set for his execution, he broke jail, and in company with one Mayfield , fled to Nevada Territory. Sheriff Blackburn , of Virginia City. Nevada Territory , was immediately informed by telegraph, of the escape of Plummer, and requested to be on the lookout for him. The sheriff immediately sent his detectives into the various towns and mining camps, in the immediate vicinity of Virginia City, to learn, if possible, of Plummer's whereabouts. Some months elapsed however, before any trace of the fugitive from justice could be obtained, but finally two suspicious looking characters made their appearance in Washoe City.

Plummer in the meantime had not only changed his dress, but had cultivated something of a pair of whiskers, contrary to his usual habit. He also wore a huge bushy wig, that hung in curls about his neck, for the purpose, no doubt, of eluding the vigilance of detectives, which he well nigh accomplished, and would have effectually done but for his frequenting gambling houses and other places of vice. The appearance of these roughs at the gambling and drinking houses of this little town, had been duly noted by Detective Williams , who kept his eye upon them until they departed for Virginia City, when he immediately notified Sheriff Blackburn of their presence in town.

Blackburn at once commenced searching the city, visiting the various haunts of iniquity, but unfortunately for him, they had been informed, upon their arrival in town, by an old acquaintance and friend of Plummer's, of the telegraph dispatch, and that Blackburn was on the lookout for Plummer, and would arrest him if he remained in town. Plummer was thereupon secreted by his friends—of whom he had many through the whole mountain country--while Mayfield strolled about town. A few days only had elapsed, when on meeting Blackburn upon the street, he roughly accosted him in this wise:

"Halloo, old chap! I understand you are the sheriff, and that you are searching for Plummer, how is it, ha?"

"Well, sir, suppose I am, or am not, what is that to you?" rejoined Blackburn, at the same instant recognizing him, from the description given by Williams, as being the comrade of Plummer.

"But I have been informed you are," continued Mayfield.

"It matters not what you have been informed," re-

Murder of Blackburn.

sponded Blackburn, "you are my prisoner," at the same time stretching forth his arm to arrest him, whereupon Mayfield drew a concealed dagger, or huge bowie knife, and uttering an awful oath, plunged it into the heart of his would-be captor, who staggeringly attempted to draw his revolver, but the unsuspected blow had been fatal, and he fell a lifeless corpse, not however until thrice pierced with the deadly weapon.

For this wilful, if not premeditated murder, Mayfield was formally tried, convicted and sentenced to suffer death upon the gallows, but previous to the day fixed for his execution, assisted by Plummer, who, in the meantime, had been lurking about, hoping to be able to render him some assistance, made his escape from the jail in which he had been confined, and in company with Plummer fled to Western Idaho.

CHAPTER III

On arriving in Western Idaho, Plummer and Mayfield found the country peculiarly adapted to their business. New mines had recently been discovered upon the Salmon, Clear Water and Boise rivers, and a large number of people had already reached the diggings.

Society was yet in a chaotic state. Gambling, robbery, and murder, were the order of the day. Years of experience had made Plummer an expert in his profession, and he now entered into his nefarious business with renewed energy. During the summer of '62, he visited nearly all the towns in Western Idaho, remaining, however, most of the time at Elk City. He and Mayfield had entered into partnership; he was now to do the gambling, while Mayfield was to look up sights. It was not long before an opportunity, in the way of robbing, was offered.

A certain fellow, who went by the name of Tom, and whose habits and character were none of the best, had accidentally fallen in with Mayfield, and finding

23

him to be of the "right stripe," as the boys say, revealed to him (Mayfield) a plan for robbing the stage coach running between Elk City and Lewiston.

He said, "I am thoroughly posted, and know positively that the coach that starts for Lewiston , on Thursday morning, next, will carry out over twenty-five thousand dollars in gold dust, and she must be stopped, and those fellows relieved of every cent of it."

"'Twont do to let her off with that sum, without making an effort to save it; but ain't there some mistake about it? How do you know that she'll have that amount aboard?"

"Why, you see, I've been watching them ar' fellers for more than a month. I've known they were making money, and they must have more than twenty-five thousand dollars."

"I say, Tom, you know the fellers."

"Well, I reckon I do; now you've hit it Jack; it's them ar' two Joneses, and that other feller by the name of Dugan, that have been mining up there on Dugan's bar. Why Jack, I've stood and seen them clean up several pounds of pure dust of a night, and I tell you, Jack, I hain't been up there looking around for nothing. I've learned all about the fellers, and know that they've got some 'ers between twenty and thirty thousand dollars, and that they start on Thursday morning for San Francisco."

"I'll be d--d if that ain't a pretty good sight, but, Tom, how do you know that they start Thursday morning?"

"Hain't I seen them buy their tickets to-day, and heard Captain Boyd , the stage agent, tell them that the coach should be ready early in the morning; but what's the use to be talking. I see you kind o' don't like the job; come, what do you say, are you in?"

"Of course I am, Tom; but you see I always like to know what I'm going into, beforehand; but I say, Tom, I've got a friend that's a bully feller; capital in these 'ere scrapes, 'specially if he gets a little crowded. What do you say, shall we want him?"

"To be sure, and we'll divide the spiles, but we'd better be off, those ar' fellers over there will think strange, our talking so long here; meet me with that ar' other chap to-morrow evening at Bob's saloon, and we'll arrange matters."

"That's a G-d d--d good boy," said Mayfield,—better known in the mines as Jack,—as he turned from the spot of this brief interview with Tom. "That arrangement will just suit Plummer," he continued. "as he hain't been having first rate luck, lately."

The beautiful September's sun was now sinking behind the western horizon, and night was throwing her sable mantle over hill and valley, as Jack hastened to inform Plummer of the proposed enterprise, and make the preliminary preparations for a temporary absence from home.

How well Plummer was pleased with the enterprise. we shall see, by observing with what energy his part of the programme, arranged the following evening. was carried out. At about seven o'clock the following evening, the three, Tom, Jack and Plummer, one after the other, dropped into the previously designated saloon, quietly passing through the bar and gambling room, and entering a private consultation room.

Bob , being "one of the boys," had arranged in the rear of his saloon, a private room, tacitly a sleeping room, but really for the accommodation of such of "the boys" as might desire secret interviews. On entering the room, Plummer found Tom and Jack alone, but engaged in low and earnest conversation. Turning upon his seat, Jack roughly introduced his friend to Tom, who, seizing his hand, at the same time giving him a piercing glance from his dark eyes, and shaking his hand cordially, turned to Jack, and observingly said, "Your friend." "Ay," rejoined Plummer, "and your friend also." For in each other they recognized old and tried friends. They passed the winter of '52 together in San Francisco, where, on more than one occasion, not only their friendship for each other, but also their bravery had been tested with the revolver and bowie knife.

Plummer and Brown,—for Brown was his real name,—now stood face to face. They had parted in San Francisco on that fatal night in February, '53. Fatal, I say, because they had made it so to one McDonald , a prominent merchant of the city, brutally murdering him in his own store, and robbing it of fifteen thousand dollars in gold.

This horrible murder will be remembered by all who are versed in the early history of San Francisco, as it aroused the indignaiton of the whole populace, and great efforts were made to apprehend and bring to justice the perpetrators of the crime. They had entered the store through a back cellar door, and finding McDonald asleep, had killed him with one stroke of a hatchet, splitting his head open, and thus avoiding any noise, they soon robbed the store of what money it contained, and quietly withdrew, without arousing any one in the neighborhood. Brown,

26

however, left the city that night, while Plummer remained some time after, but finally going to Sacramento, and subsequently to Nevada City, as before stated; but neither of the parties had seen the other since that bloody night, until the present meeting in Bob's saloon.

After warm congratulations, Plummer and Brown seated themselves with Jack, around a table, from which a tallow candle was diffusing its dim light through the darkness of the room, which, from its general appearance resembled more a receptacle of the dead, than it did of the living. In one corner, upon a low bench, were situated a half dozen human skulls, of both the "simon pure" natives, and the Anglo Saxon races, while upon the opposite wall hung a number of Indian scalps, all of which, in the somberness of the room, had a ghastly appearance, and as the flickering rays of light fell upon the broken and distorted skulls, one seemed in the presence of so many hissing devils, whose glaring eyes and disfigured countenances were making grimaces, and gnashing their teeth, as Brown seated himself upon the end of the bench upon which they were located, slightly disturbing them, at the same time, apparently whispering out upon the thickness of the evening, the horrid word, "murder," which should only be uttered in the darkness of the night, and in the presence of devils.

This was indeed a most befitting place for the consumation of such arrangements as were to be made that evening, and although there was something in the locality that sent a paralytic thrill through the whole physical frame, and stifled the utterance for the time being, yet there was also an indescribable something

27

—term it influence if you please—created from the analogy of the peculiar surroundings, with the hellish work to be accomplished, that gave nerve to the desperadoes, who, after drinking each other's health, proceeded to lay their plans for robbing the stage the following morning, which worked admirably. Echo Canon was the place selected for the robbery, it being nine miles by the circuitous route the stage had to travel, while it was really only five miles on an air line, from the town. The coach was to start at three o'clock in the morning. They were to disperse, and meet again at three o'clock at the appointed canon, and await the coming of the stage, commit the robbery, bury the money, and return over the mountains to the city, before the news of the robbery could reach town, and if necessary to more effectually hide their guilt, join in the search for the robbers.

The time and place, as well as the men who were to execute the work, were most admirably adapted to the occasion, and as the coach came rattling down the canon, it was daringly stopped and robbed, without firing of a gun, its inmates seeming glad to get off thus easy. The money, amounting to some twenty-seven thousand dollars, was buried, and the robbers back in town before the coach had arrived on its return.

Upon its arrival, however, great excitement prevailed, and every exertion possible, was made to ferret out and bring to justice the robbers, in which the guilty parties took an active part. After weeks of fruitless effort to obtain any trace of the guilty ones, the excitement quieted down. The first part of their programme—relative to robbing the stage—had been accomplished much easier than they had expected, for indeed they had supposed, and in fact had prepared for a desperate struggle, and now that they had succeeded so well in hiding their guilt, by the feigned efforts they had made in assisting in the attempt to hunt down and ferret out the perpetrators of the crime, all that now remained for them to do was to resurrect their money, and properly dispose of it without creating any suspicion.

In the meantime, Plummer had secretly—in his own mind, laid a plan to escape with his share of the funds, as soon as an opportunity offered, nor was he going to wait long for an opportunity, for his fruitful mind had already conceived a plan which, if accepted by the secret Police Committee, would fully enable him to accomplish his design.

His plan was to make a trip to Portland, in Oregon, under the guise of a secret detective, for the pretended purpose of apprehending the robbers, but really for the sole purpose of shipping his money to the States, after which he was to return to the city, and report a fruitless search. The ability which his natural shrewdness, and the experience that he had had at Nevada City, while marshal, had enabled him to display, as a detective, during the excitement, had in a degree secured him the confidence of the better portion of community. His plans were, therefore, readily acceded to by the secret committee, and the details of his trip down the Columbia arranged. He was now provided with a horse and saddle, preferring to go as far as Lewiston on horseback, and start in the night, as he said, to going by stage, which, as he suggested to the committee, might create suspicions in the minds of some lurking friend of the robbers, who would by some hook or crook, inform them of his (Plummer's) departure, thus putting them on their guard.

This was, indeed, a happy thought for Plummer, as it not only blinded the eyes of the committee more effectually as to his guilt, but also enabled him to escape with his money, which he could not have done by stage, without being detected.

During the consummation with the committee, of their arrangements, Plummer and his accomplices had quietly stolen, at dead of night, over the mountains, and raised and divided their ill gotten money. Plummer, however, took good care to privily secure his portion in a mountain gorge, among the broken and distorted rock, until he should come that way, while on the road to Lewiston.

His arrangements were now completed, and at twelve o'clock, on the night of the 20th of October, he stole noiselessly out of town, unknown to any one except his accomplices, and a few of the secret committee. On reaching Lewiston he took passage on a steamer for Portland , where he spent some two weeks, during which time he shipped ten thousand dollars to the States, which he had accumulated by gambling and robbing, while in Idaho. Leisurely returning, he stopped a few days at some of the most important towns, making such inquiries regarding the robbery as he thought expedient, of the authorities, finally reaching home at Elk City, after an absence of six or eight weeks.

The Robbers Dividing Their Money.

CHAPTER IV

During Plummer's stay in Western Idaho, he formed the acquaintance of a Miss D. , whose real name we omit, as he subsequently made her his wife. It is due to her to say, that she was a lady of both refinement and intelligence, as also of respectable parentage. She evidently had not been familiar with his character before marriage, or she in all probability would not have given him her hand in marriage. It was during his trip as "secret detective" that he had first met her, and by misrepresentations, she was led to the belief that he was a man of respectability as well as standing at home. She was, however, soon to move over the mountains to Sun River, and an arrangement was made between her and Plummer, that he should meet her there the following spring or summer.

On his return to Elk City, he found all quiet, so far as any excitement relative to the late robbery was concerned. Reporting a fruitless search, he soon joined his old accomplices, and the long winter nights

33

were wiled away in drinking and gambling, little of consequence occuring in the way of robbery or murder for some time subsequent to the notorious robbery of the stage.

A few shots had occasionally been exchanged over a disputed game at cards, and one man, who went by the name of Mike, had been stabbed with a bowie knife, but not fatally. About the first of February, a row took place between Tom and one Dening , at Bob's drinking saloon, in which Tom severely wounded Dening, by a shot from his revolver. No notice was, however, taken of it by the citizens, because it had not proved fatal. A rumor was soon in circulation that Dening's physician had pronounced the wounds not fatal, and that he was slowly recovering. In the meantime Dening had said to his friends, who called upon him, "that he would follow Tom to hell, but what he would kill him, that is, if he (Dening) should recover; and that he would shoot Tom on sight," etc. Tom, of course, heard of these threats, and knew that his safety would be doubtful if Dening was allowed to recover. Thereupon, he counseled Plummer and Jack, who advised him to put Dening out of the way if possible. Dening was at that time occupying a room in the second story of a building, and was fast recovering.

Tom preparing himself with the proper weapons. stole noiselessly and fiendish-like up the outer stairway, followed by Plummer, who waited at the door while Tom entered the room, where he found Dening asleep, when, with one stroke of his bowie knife, he cut Dening's throat from ear to ear, without creating any alarm, and making their escape unseen, except by a citizen, whom they met while turning the corner of the street a few rods from the house, where the horrible crime had been committed.

The murdered man was soon found with his throat horribly cut, and the bed upon which he was lying, saturated with blood, while a large pool had accumulated upon the floor. The alarm was given, and the excitement became intense. A large crowd of miners and business men gathered around the house where the helpless man had been brutally butchered, all clamorous to deal swift justice to the guilty.

Plummer and Tom were soon traced to their cabin by the citizen, whom they had unfortunately (for them) met, and who happened to know them. They were dragged from their beds, having hastily retired, and their clothes and weapons examined. Blood was found upon Tom's knife and sheath. His hands were then examined, and found stained with fresh gore. Plummer was then examined, but no trace of blood could be found either upon his weapons or person. The evidence of guilt upon the part of Tom was sufficient to condemn him in the eyes of an excited populace, and he was immediately dragged off to a signpost and hung, not however until he had confessed his crime, stating his reasons for doing so to be, that he apprehended Dening would kill him, if he should recover, and that he considered, from the threats Dening had made, that he (Tom) had killed him in self defense.

Plummer not being implicated in the matter, farther than having been seen with Tom, was released the following day, under the solemn injunction to leave the country within twenty-four hours, or suffer death.

Acting upon the principle that a "word to the wise is sufficient," he immediately left the country accompanied by Jack .

He did not return to California or Nevada, for he was too widely known along the Pacific, but on the contrary, sought out the most remote settlement, where in the absence of civilization, with its accompanying laws and safe guards to persons and property, he could enter anew upon his hellish work of robbery and destruction of human life. That locality he found in Bannock City, east of the Rocky Mountains . Arriving there in the spring of 1863, after a long and wearisome journey, over the snow-clad mountains, in which he and his comrade suffered much from cold and hunger.

Finding Bannock, at that time, ruled by a set of lawless devils, he thought it advisable to make a bold debut upon his arrival in the city. Consequently while drinking with Jack in a saloon the very day of their arrival, he deliberately shot him (Jack) without there having a word of altercation passed between them. The ball took effect in the chest, but not killing him instantly, but for which circumstance much of the foregoing narrative or biography of Plummer would have been wanting, as from him we receive many of the foregoing facts in his dying hours.

He also informs us that the probable cause of Plummer's shooting him was in consequence of some little matter of a private nature, connected with a woman, about which they had previously had some words.

Thus Plummer, in cool blood, had murdered a comrade who had stood by him for years, and on one occasion had imbrued his own hands in human blood, for the sole purpose of preventing the arrest of Plummer as before stated.

CHAPTER V

Spring having at length arrived, the people of Bannock set themselves at work for the purpose of establishing civil law, hoping thereby to bring order out of disorder, and to be able more effectually to resist the desperadoes, who by this time had become very numerous in that locality. So much so, indeed, as to render it quite unsafe for business men,—especially, if they were known to have money,—to travel the streets in the day time, and much more unsafe to be out of doors at night.

An election was accordingly called, for the purpose of electing a judge and sheriff, which resulted in the election of one Burchett to the judgeship, and Henry Plummer, of California notoriety, to the sheriffalty. Through his political chicanery,—for he had learned the arts of the politician while at Nevada City—he had secured the nomination from the "bummer" party, which was at this time apparently in the majority. Apparently so, from the fact that they had made it dangerous for a person to oppose them in their hellish work of ranting and brawling about town, brandishing revolvers and dirks, and occasionally killing a man. To oppose them, was a signal for a knock down, or the immediate assassination of him who had the audacity to intercept their plans.

This state of things kept the masses of the people, who, as yet, had no organization, terror stricken, while the "bummers,"—as they are termed—had it all their own way. Plummer having acquitted himself so nobly upon his arrival in town, by killing his comrade in cool blood, as previously stated, had not only merited their especial favor, but had also given himself notoriety, which had the effect to draw around him the entire multitude of ruffians of the country, and as a consequence he received the election.

Plummer having obtained position, was, of course, enabled to prosecute his devilish work under an official garb. He had eagerly sought to become sheriff, knowing that by occupying that position, he could materially aid any of "the boys" who might be detected in any of their lawless operations, and not only that, it would also place him where he might commit the most heinous crimes, and go unpunished.

The election being over, Plummer made arrangements for an absence of a few weeks, with a view to going to Sun River, for the purpose of marrying Miss D., of whom we have before made mention. Meantime, a trader by the name of Granger , from the West Side, who had come over the mountains with a pack train, was closing out his goods for the purpose of immediately returning home.

Plummer having learned of him, had "spotted him"

in his own mind, as the saying is, and determined by some hook or crook, to relieve him of the proceeds of his goods, which he had learned would amount to several thousand dollars. This was a snug little prize, and it would not do to let a golden opportunity pass. Fortunately for Plummer, as circumstances would have it, Granger called upon him, when the following interview took place.

After passing the usual interchange of Western courtesies, Granger said, "Mr. Sheriff, I called for the purpose of having a few words of consultation with you, and to ask your advice in certain matters."

"I shall be pleased to assist you in any way in my power, either officially or otherwise," rejoined Plummer.

"I have just arrived from Lewiston, with a pack train of mules, and have sold out my stock of goods, and desire to procure the services of one or two good men to assist me in returning with my train, over the mountains, as the men who came with me, are not going to return, I having engaged them only for the purpose of assisting me through to this place, they preferring to remain in the mines; and it having been suggested to me that you would be the most likely of any one, to know of parties, if there are any, contemplating a trip over the mountains, and if not, of some one whom I could hire to accompany me, therefore concluded to call and see you with regard to the matter."

"I really don't know of any one just at present, but had you been here a week earlier, you might have got off with a train of five wagons that left here last week for Walla Walla, but call in again this evening, and in the mean time I will make some inquiries, and perhaps can be of some service to you."

As with a flash of lightning, the whole scheme of assassination and robbery passed before Plummer's fruitful imagination, and as Granger withdrew from the office, Plummer paced the floor with unusual celerity, with his head slightly inclined, evidently in deep thought. The theme was indeed a novel one.

He was about to enter the connubial state, and the idea of passing the threshold of the bridal chamber, and presenting himself to her, whom he knew to be the embodiment of innocence and virtue, all besmeared with human gore, fresh from the bleeding wounds of a murdered man, was shocking, even to his demoralized and blunted sensibilities. To ruminate in thought, to the innocent, is a pleasure, but to the life-long murderer, however bright his skies may appear, or with whatever prospects his future may open, thought, deep and unrelenting thought, is like a serpent coiled within, and ever stingeth, even unto death, like an adder. He had not, as yet, forgotten the once loved, and angelic Miss B., of Nevada City, before whose shrine he had worshiped with all the devotedness of a true lover, and from whence he had been spurned, only by the intervention of his own crime.

Now that he was again betrothed, to commit any crime that should re-enact those early scenes, and bring about a like result, would simply be, to darken his whole future, and cast a withering blight upon his domestic prospects, for he loved the lady to whom he was now engaged, with all the intenseness of a passionate heart, and although it had become as

42

adamant to all else, yet, from its depths welled up an intense glow of kindred love, to that he had borne for his kind and devoted mother, in days gone by.

Starting as from a dream, and raising himself at full length, with an air of defiance and determination, he exclaimed, "I will execute my plans, let the consequences be what they may, my will shall become the law. A man is not a man, who cannot rise to the dignity of the occasion;" so saying, he left the room.

During the day, he made it his especial business to call upon a fellow who had just arrived from Nevada Territory, and who went by the name of Texas . Plummer had known him while in California, and knew him to be of the right stripe. After a brief interview, they parted. What the programme fixed upon was, subsequent events will show. At seven o'clock that evening, Granger called at Plummer's office, as previously agreed upon. Plummer welcoming him, said he was sorry to inform him that he had been unable to learn of any one who was contemplating a trip over the mountains at present, but he knew a very reliable man, whom he had no doubt might be got to accompany him. Just at this juncture Texas entered the office, when Plummer said, "Ah! it's you, is it? I was just speaking of you to this gentleman, as being a suitable person, from your experience with pack trains in California, to assist him in taking his train over to Lewiston; that is, provided he could induce you to make the trip."

"I reckon I might, provided I can dispose of a few traps that I have, which I had better dispose of if I go, as I should not return for some time."

"Very well," said Plummer, "if you and Mr. Gran-

43

ger can make the arrangement, I will assist you in selling your goods," then turning to Granger he continued "with his assistance, I think you will make the trip in safety, and as I am about starting for Sun River on business, I will consent to accompany you as far as Deer Lodge, myself."

Thanking him for his kindness, Granger turned to Texas, and soon completed the arrangements, and upon the morning of the following day but one, the three were seen winding their way up the rugged cliffs that environ the little town of Bannock, and bearing away to the northwest.

The evening of the second day found them at Bighole Pass . Here they left the main trail, and encamped near a spring, in a deep canon, about a mile from the road. After enjoying a hastily prepared supper, the three seated themselves in front of a blazing camp fire, and wiled away the evening hour at a game of euchre. It was a lovely night in the month of May. The sky was cloudless, and the silvery rays of the moon were falling upon mountain and glen, as the evening twilight disappeared. All was as hush as the midnight chamber of death, save the low whispering of the fragrant mountain zephyrs, as they rustled through the leafless willows that hedged the neighboring brook.

Time sped on. The hour of eleven had arrived, and the three retired for the night. Granger, unsuspecting of danger, soon fell asleep. Plummer and Texas, whose plan was to assassinate him while in the act of sleeping, now quietly arose; Texas plunging his bowie knife into the breast of his victim, while Plummer, with a huge dirk, nearly severed his head from his body, with one stroke across the throat. An instant more, and he was a lifeless corpse, prostrate at their feet, without having had an intimation of danger, or the slightest opportunity to defend himself. They at once stripped him of his clothing, taking from his person a belt containing three thousand dollars in dust, and dragging his body some forty rods from the camp, and burying it beneath a mass of rocks, after which they took four thousand from his valise. They then burned all his clothing, pack saddles, and every combustible portion of his train. Carefully gathering from the remains, all the pieces of iron, bridle, bits, etc., and threw them into a deep hole in the creek, lest they might be discovered, and lead to the detection of their crime.

The mules were then stampeded in every direction, so that if found, but one or two would be together, which they thought would not excite suspicion; as to find an odd mule or horse was a common occurrence in this mountain region. Morning having arrived, they mounted their steeds, each having his share of the profits of the night, and started. Plummer taking the trail leading to Fort Benton, while Texas started for the Bitter Root Valley, where he was subsequently hung by the Vigilance Committee; but not, however, until he made a confession, a part of which we have just related.

Plummer soon reached Sun River, where he found his intended living with a brother, at a "fur trading post." We will not detail the minutiae of the wedding, but simply say that he was immediately joined in wedlock, and soon on his back trail for Bannock City, where he lived happily with his wife, until the first of October, when she left for the States, on a visit.

CHAPTER VI

During the month of June, 1863, gold mines were discovered on Alder Creek, seventy miles northeast of Bannock City. At this place Virginia City sprang into existence, almost instantaneously; people emigrating thither from all settlements within a radius of hundreds of miles. About the last of June, Buck Stinson, Hays Lyons, and Charles Forbes, three noted roughs, had "spotted," (as the saying is,) an old miner, whom they knew to have been making money at Bannock; but who had sold his claims, and was going to the Virginia City mines. It was known that he had taken out some thousands of dollars in dust; and acordingly, they laid their plans to go over there and rob him when he should get quietly settled down, knowing that he was going to locate in rather a remote part of the gulch, where they could easily accomplish their work.

Sheriff D. , who was deputy under Plummer, was a man of reputed integrity, and had received his

47

appointment from Plummer, through the earnest request of the better portion of community. Being informed by the timely intimation of a friend, that some scheme was on foot for either the robbery, or murder, of a certain man, by the boys, he therefore immediately set himself to work to ferret out, and, if possible, to thwart their plans.

Approaching Stinson and Lyons in their cabins, he said, after pleasantly passing the time of day,

"Boys, are we alone?"

"Certainly; what have you to say?" rejoined Stinson.

"Well, nothing in particular. I just thought I would drop in as I was passing by, and see how you were all getting along. Any thing new? I suppose Old California has got quietly settled down over there by this time. Wonder if he took all his dust with him?"

"I suppose he did, as he would not be fool enough to leave it with any one here."

"I have been thinking," said the deputy, stopping, and again asking if it were possible to be heard by any one, "that he ought to be relieved of that dust, and I have a plan by which it can be accomplished; and knowing that you were 'all right,' thought I would give you a chance in, if you would like it. What do you say, boys, would you like to have a chance in?"

Stinson, not being satisfied that the deputy was "one of the boys," said, "why do you come to us? We had thought you were too honest for any thing of that kind."

"But," said Lyons, "he wears the cordon knot."

"What means that?" continued Stinson.

"Eternal friendship!" replied the deputy.

He had seen the peculiar knot in the neck-tie, worn by Plummer, and other noted characters; and on one occasion heard Plummer telling a rough, that the knot signifies friendship.

"That will do," said Stinson. "I see you are all right. We have made our arrangements to see Old California next Sunday ourselves. But if you wish a hand in, we should not object to having another chap along; and if you will go, we will make a joint thing of it."

"Very well; I am with you," said the deputy.

"Meet us at Virginia City, Sunday morning, at eight o'clock. We shall be there in readiness," continued Stinson.

Entire secrecy was enjoined upon all, and the parties dispersed to meet as above indicated. The deputy's object being only to learn of the plottings and designs of these notorious ruffians, immediately went to Virginia City, and secretly notified California of their purpose to rob him, whereupon he immediately moved into the city, for the purpose of safety.

The designated Sunday at length came, and with it the three desparadoes, who, on reaching the city, soon learned that the deputy had been there several days, and that California in the meantime had moved into town. They at once came to the conclusion that the deputy had betrayed them, and the fact that he had not met them that morning, according to agreement, was conclusive evidence to their minds of his duplicity in the matter.

On learning of his whereabouts, they immediately repaired to the house. Stinson entering, asked the deputy to step to the door, as he wished to speak with him a moment. Unsuspectingly, he followed him out of the house, and partially around the corner, when Stinson exclaimed, "You d--d scoundrel, you have betrayed us, and we will now have revenge. There! take that you d--d son of a b-," at the same instant, both Stinson and Lyons leveled their revolvers at his breast and fired, the balls taking effect in his body. The third man, Forbes, had retained his fire, so that in case they did not make a good job of it, he should be ready to finish the work, but as their victim fell apparently lifeless, Forbes did not fire. The deputy was immediately attended by physicians, but expired in a few hours.

An immense crowd of people at once assembled, this being the first murder committed at Virginia City, and means were adopted for the arrest and trial of the guilty parties. The villains were accordingly apprehended and committed to prison, to await their trial. Much excitement prevailed, and it was with difficulty that the crowd was kept from breaking into the building, in which they were confined, and lynching them, without any form of trial; but through the efforts of a few leading men, quiet was preserved, and a committee, consisting of several, were selected for the purpose of sitting as a court, to hear the case, and decide what punishment should be meted out to them.

After having heard the case fully investigated, the court unanimously agreed as to the guilt of Stinson and Lyons, and hence decided that they should be hung, while Forbes should be banished from the country. Their decision was read and referred to the assemblage, which, by a vote, concurred in the decision of the court. Sentence was then pronounced by the court; the time fixed for the execution was between the hours of nine and eleven, the following morning.

In the meantime an officer was appointed by the court to take charge of the execution, who immediately remanded the felons to jail, stationing a strong guard about the house, and ordering a gallows to be erected, coffins made and graves dug, which were all in readiness at the proper time, the following morning. During the night demonstrations were made by members of the gang, for the purpose of rescuing the condemned, but finding so large a force on guard, no efforts were made further than threats. Morning at length arrived, and as the hour of the execution drew nigh, an immense number of people congregated about the jail, and when the "death wagon," which contained their coffins, and drawn by one horse, came around to the door of the jail, the criminals came out and very deliberately seated themselves upon their coffins. The crowd gave way in front, and the wagon moved forward followed by the multitude, whose steady tread only echoed through the stillness of the morning, which was as hush as the grave.

On arriving at the place of execution, one Judge Smith , a gambler and lawyer, formerly of California, but more recently of Colorado Territory, arose and standing upon the "death wagon" with the murderers, begged the crowd to listen to him a few moments, before executing the boys, which was tacitly granted, there being no objection made.

The judge, who was more noted for his eloquence than for his legal ability, now found an opportunity to pour forth his oratory, which he most effectually did in behalf of the criminals, who were standing apparently upon the brink of eternity Turning to

Lyons, he said—"that while that young man was an infant, he had dandled him upon his knee, he had watched his course from thence to manhood, and knew him to be a young man of promise; and notwithstanding he had erred, he (Smith) prayed that they would revoke the death sentence and banish them instead."

After a speech of at least an hour in length, he read a letter written by Lyons to his mother while in jail, and also one by Stinson. These letters were quite affecting, but the speech of Smith had been an effort upon the brink of the grave, an occasion of all others which stirs an orator from the depths of his soul, and his effort had aroused the inmost passions of his auditors and awakened a sympathy in the most hardened heart. Women gave vent to their feelings in most bitter sobbings, while men forgetful of their manhood, stood awed before the speaker; and when the vote was re-taken, the sentence was revoked and the "scape gallows" given twenty-four hours to leave the country. Thus justice was defeated, and the gallows and the grave alike robbed of their just due.

> "The greatest attribute of heaven is mercy;
> And 'tis the crown of justice, and the glory,
> Where it may kill with right to save with mercy."
> "Mercy is not itself, that oft looks so;
> Pardon is still the nurse of second woe."

52

Great Speech of Smith in Behalf of the
Condemned.

CHAPTER VII

It had long since become apparent, from the concert of action upon the part of the desperadoes, that a secret organization existed, which was governed and directed by some general head, and while all had their opinions as to who its leader was, yet none could positively tell.

It was enough to know that the country was infested with heartless desperadoes, who were robbing and murdering in every valley and mountain gorge, where an opportunity was offered, or an individual could be found. Men were daily starting for the States, or moving from one locality to another, which offered ample opportunities for the prosecution of their nefarious work.

Under this state of things, it would not be strange that men should mysteriously and suddenly disappear, and never be heard of again. Indeed, it became of such common occurrence to hear of a robbery or murder, during the summer and autumn of '63, that little or no attention was given to any thing of the

55

kind, unless it occurred in town. A murder was, however, perpetrated on the morning of the 19th of August, in Bannock, which was so unprovoked, and so outrageous in its detail, that it aroused a feeling of indignation, and a desire for revenge in the mind of the whole community. An old man, whose head was silvered with the frost of many winters, and whose character was spotless, beloved and respected by all who knew him, and without an enemy in the country, had been brutally assassinated in his own cabin, by one Peter Horen. The assassin had gone to the cabin of his victim, who, by the way, was one Lawrence Kealey, formerly of Davenport, Iowa, at an unusual early hour in the morning, and deliberately assassinated him. It appeared from the evidence given at the trial, that Horen had formerly been a partner of Kealey and a Dutchman, in mining operations, and that some little difficulty ocurred between them.

The Dutchman, however, testified that Horen was the aggressor, and that Kealey had always treated him with kindness. Horen, however, determined to dissolve the partnership business, and accordingly went to Plummer, who was at that time sheriff, and consulted him as to what he should do in the premises. What the advice given at this interview was, we shall take occasion to relate hereafter. Sufficient for the present, is it to know, that after a long consultation, the parties separated, not, however, until Horen had placed in the sheriff's hands his mining interests, and also a number of shares in a Ditch Company, for the purpose of having them sold. Plummer immediately advertised the property to be sold the following Friday.

56

Shooting of Kealey by Horen.

In the mean time, Horen purchased a horse and saddle, hiring the horse kept at a corral in town. He also purchased a navy revolver,· having one at the same time that he had brought into the country. He had evidently made Kealey the object of his pent-up malice, and determined upon glutting his vengeance, by murdering him.

The day before the horrible occurrence, he was seen in a valley west of the town, practicing target shooting with his revolvers. On Friday his property was sold, bringing about five hundred dollars in gold, which was paid to him late in the evening, by Plummer. At five o'clock the following morning, he rapped at the door of Kealey's cabin, who had not yet risen from bed, but upon hearing Horen's voice, unsuspectingly opened the door, when Horen, appearing with a revolver in each hand, deliberately leveling one at Kealey and firing, the ball taking effect in the left breast. Kealey instantly closed the door, with the cry of murder. Horen then stepped back from the door and discharged two more shots, the balls passing through the door, but missing Kealey, who, in the meantime, had fallen upon the floor. Horen, after fruitless attempts to force a passage between the logs of the cabin, for the purpose of firing through to finish his work of death, again went round to the door, and opening it, fired at the prostrate body, which was already weltering in blood, and writhing in the agonies of death. lodging a second ball in the left breast, either of which would have proved fatal.

Satisfied that he had accomplished the work, he hastily repaired to the corral, where he had previously saddled and bridled his horse, and was just in the act of mounting, when arrested by the sheriff, and taken to a place of confinement, and put under guard. A coroner's inquest was immediately called, after which the remains of Kealey were buried in due form. A trial was also immediately instituted, in which counsel were employed upon the part of the people, and also for the defense of the criminal. The case having been fully investigated, the guilt of Horen was so apparent, although his attorney endeavored to make a case of monomania, that the verdict of the jury was unanimous as to his guilt, and the court prĕnounced the sentence of death, giving him only a few hours to live, after the verdict was rendered.

It now became the duty of Plummer to execute him, and he accordingly had a gallows erected, and at half past nine o'clock in the morning, which was but two hours subsequent to the rendering of the verdict, a wagon drawn by one mule, came around to the door of the jail, when Horen was assisted in getting into it, he not being conscious of what was going on, from the effect of drugs. He was thus dragged off to the gallows and hung, being at the time perfectly insensible. Why he should have been thus drugged, so as to be unable to speak, and much less to converse with his friends, was a query in the minds of many who witnessed the execution. That there was some reason for it, all admitted, but supposing it to be for the prisoner's benefit; perhaps for the purpose of removing the apprehensions of death, and to quietly let him down into the dark valley, unconscious of the transition, thus it was considered, and passed quietly by, and soon forgotten.

Recent events developed the true reason for his

being stupefied with drugs. Long John, who turned state's evidence, and revealed the secrets of the banditti, states that in a secret conversation that he had with Plummer, shortly after the execution of Horen, that Plummer informed him that when Horen first consulted him with regard to Kealey, and his interests in the partnership claims, that he (Plummer) advised Horen to do just as he had done with regard to killing Kealey, and that he agreed to assist him in making his escape, if arrested. He also stated that his reasons for giving this advice were, that he expected the murder to have been committed a few days sooner, which would have left Horen's property in his hands, which he, at the time of giving the advice, supposed would sell for much more than it had, hoping to be able at the same time, through his official duplicity, to assist Horen in his escape, which he said he regretted very much that he had been unable to do, and therefore he had made Horen believe up to the rendering of the verdict, that he would not be convicted, and that if he was, that he (Plummer) would give him an opportunity to escape, but the immense excitement the murder had created, rendered it impossible, and he saw that something must be done to close the mouth of Horen, lest he should make a confession which would involve him, which he boastingly said he had accomplished with drugs, without creating the slightest suspicion on the part of the people.

CHAPTER VIII

ROBBING A STAGE COACH.

As a prelude to the robbing of the stage, running between Bannock and Virginia City, which took place on the morning of the 18th of October, 1863, we would say that the parties robbed were not the ones the ruffians, or rather "road agents," as they are called, intended to rob when they set out on their mission from Virginia City.

They had "spotted" one Captain Fisk , of the United States Army, who had escorted an emigrant train across the country from St. Paul that season, and who had just sold out his government train and was about starting for the States. His train, consisting of horses, mules, wagons, harnesses, saddles, and a variety of camping utensils, had been advertised and sold at public auction, which had, of course, given the "road agents" an opportunity to learn the probable amount of money the captain would have, which was thought would reach several thousand.

The captain, after having settled up all his business matters, left Virginia City on the morning of the 16th.

61

The time fixed for his departure having previously
become known to "the boys," who are ever on the
alert looking out for golden opportunities; accordingly
the night previous to his departure, a noted gambler
who went by the name of "Dutch John ," quietly left
the city, accompanied by another fellow, for the
purpose of robbing the captain.

Whether or not Plummer, the sheriff of Bannock,
was connected with this robbery, is not known, but
it is fair to presume that he was, from the fact he was
at Virginia City at the time, and had made himself
quite familiar with the captain during his stay there.
This fact, taken in connection with subsequent developments, which proved the existence of an organized
band, with Plummer at its head as captain, in which
all the parties connected with the robbery were active
members, and all together with Plummer finally suf-
fering death upon the gallows at the hands of the
Vigilance Committee, would seem to indicate complicity in the matter upon his part, if indeed he did
not instigate the whole scheme.

On the arrival of the stage at Beaver Head Station .
the robbers were seen upon the bank of the river
about a half mile distant. The stage turned up to the
house for the purpose of obtaining some refreshments
and changing horses. During the short stay of the
coach, Dutch John (one of the robbers,) came down
to the ranch, and after drinking with the captain and
others, he took a scrutinizing glance at the stage, for
the purpose of ascertaining how well the party was
armed, at the same time examining the captain's
carbine and other weapons, which he found in abundance for almost any emergency, making some remark

to the captain with regard to his carbine. The captain replied, "I have got you 'spotted,' old chap, and have brought that carbine along for your especial benefit."

It is evident that his examination convinced him, that any effort upon their part to rob the stage would not only prove fatal, but would be a very hazardous undertaking, as there were six passengers beside the driver, and all were armed with revolvers and guns of some kind, and hence they did not make the attack, though passing and repassing the stage several times during the day, and finally disappeared in the bluff, just before reaching Rattle Snake Ranch. At this ranch, which is a stage station, fifteen miles from Bannock, situated upon the bank of Rattle Snake Creek, the stage put up for the night, but reaching Bannock the following day in safety.

Disappointed in their enterprise, the ruffians determined not to give it up so, and therefore lurked among the hills awaiting the arrival of the next coach, which they knew would be but a day later. The morning of the 18th found them in Spring Gulch; about eight miles from Bannock. This was a well selected spot for the perpetration of any crime. A deep ravine walled in with irregular rocky bluffs, some of which overhung the road in its stupendous meanderings.

Upon seeing the stage in the distance, they secreted themselves behind the point of a bluff, where an angle in the road would bring the stage within a few feet of where they stood. Mounted upon their steeds, with double barrel shot guns in their hands, and masks upon their faces, they awaited the approaching stage, which soon came thundering down the gulch at full speed; rounding the point of the bluff, the crack of the driver's whip enlivening the pace of his mustangs, while the presence of a merry bottle just being circulated within the stage, was the occasion of much merriment, giving vent to the melodies of a mountain song, in which all were joined. Thus the coach unsuspectingly glided along, until it reached the desired point, when with a bold dash, the two robbers stood beside the stage, mounted upon their steeds, one with his gun aimed at the driver, and the other with a like weapon aimed at the passengers. This, of course brought the stage to a stand. Dutch John, with an awful oath, accompanied with severe threats, then ordered all to throw up their hands. This order was given, of course, to prevent their being able to use their weapons; and was immediately complied with. Bill Bunton , one of the passengers, threw up his hands, with the cry, "Don't shoot! don't shoot! we will give up our money." This exhibition of fear, by one of the passengers, had the effect to intimidate the rest.

"Get out of the stage, you G- d- roughs," continued the robber, at the same time ordering all to continue to hold up their hands, under the penalty of having their heads blown off if they should make the slightest attempt towards resistance.

Bunton was the first to reach the ground. He having no weapons was then ordered to hold the horses, and the driver invited with a threat to alight from his box, after which he was compelled to disarm the passengers, one after the other, throwing their arms into a pile, at a reasonable distance from the stage. He was then forced, with a gun still leveled at his

Stage Robbery between Bannock and
Virginia Cities.

head, to search the pockets of each, and deposit the money thus obtained, with the weapons.

After all their pockets had been thoroughly searched, the robber turned to one McFadden , from whose pockets they had already taken two small purses, containing three hundred dollars in dust, and exclaiming, "Off with that belt, you d- son of a b-; you are the man we are after;" at the same time ordering the driver to tear open his (McFadden's) shirt bosom, and take off his belt that was encircled next to his person. McFadden hesitated, and the driver quibbled as to tearing open a man's bosom, when with a deadly aim, with both barrels cocked, the robber sternly commanded, and in an instant more, McFadden was stripped to the skin; a large belt containing three thousand dollars in dust was taken from his person, and placed with the rest of the trophies.

All having now been thoroughly examined, Dutch John, with a repetition of the latter oath, ordered all into the stage. Turning to the driver, he continued, "You d- scoundrel, drive off at the top of your speed." All were apparently glad to get off without the loss of blood, although there were four passengers besides the driver, and all were armed except Bunton; some of them having two revolvers, and others having one gun and a revolver. The two robbers continued mounted upon their horses until the stage disappeared beyond the bluffs in the distance, when they secured their money and decamped.

The stage rolled away in the distance, soon reaching Bannock, where the passengers related the tale of their misfortune, in which Bunton took an ac

66

describing the robbers, and relating many of the particulars connected with the robbery. This was the first stage robbery in Eastern Idaho, consequently much excitement followed, with a vigorous effort upon the part of the law abiding men of the country to apprehend the guilty ones, which proved unsuccessful, from the fact that the very officers whom the people expected to ferret out and bring to justice the perpetrators of the crime, had instigated and planned the whole scheme, and now simply connived with the robbers. After events, however, throw some light upon the matter. Bunton, some months subsequently, was arrested at Deer Lodge, by the Vigilance Committee, and hung, but before his execution he confesed his complicity with the robbery of the stage, stating "that it was their intention to have robbed Captain Fisk, but finding him and his party too well armed, concluded to let them pass, and rob the next coach. He further states that he ascertained, by careful observation, the night previous to the robbery, that McFadden had a concealed belt, and that he supposed it contained much more than it did, as Mc had been working a very rich claim, and after making these observations, he had communicated the facts to Dutch John, who came to the ranch at a late hour of the night, when all were asleep except Bunton, for the purpose of learning the result of his observations." He also confessed that the night previous after making a careful examination of Captain Fisk's party, he had advised the robbers to let them pass, as he did not think they could be robbed, and that on the morning of the robbery, he had taken passage in the stage to Bannock, for the sole purpose of intimidating the passengers, as he had done, by crying for mercy, hoping thereby to assist in the robbery, which, he said, was in accordance with his previous arrangements with Dutch John.

In order that the reader may more fully understand the sequel to this robbery, it is proper to say, that this ranch was kept by a man by the name of Pickett, who was a member of the banditti, and that William Bunton was a sort of clerk or bar tender for him. In this rural district, where hotel or ranch keeping is conducted, in the primitive style, each guest carrying his blanket or bedding, and all retiring promiscuously, as the case may be, upon the floor of the bar-room, which is usually the only room in the house, except the kitchen. Especially was this the case at this ranch, which of course, gave Bunton ample opportunity, to not only become posted as to the amount of money, arms, etc., in the possession of those stopping at the ranch over night, but also to communicate the same to others of the gang, who were lurking among the surrounding hills, ready to embark in any enterprise, so as it only promised a goodly harvest.

CHAPTER IX

MURDER OF A DUTCHMAN BY IVES.

The summer and autumn of '63 had now well nigh passed, during which time Virginia, Nevada and Summit cities sprang up , situated in Alder Gulch, perhaps better known as Stink Water Mines. The richness of these mines had created a fury of excitement through the surrounding country, and consequently a concentration at these cities and mines, of miners, and business men of every description, merchants, mechanics, lawyers, doctors, school teachers, clergymen and desperadoes were all there.

The latter, like the vulture, were infesting every nook, crook and by-path of the country, where an unsuspecting victim would be likely to be found, ready to pounce upon, and rob, or murder, as the case required, any who might chance to come within their reach. Here the desperadoes from Colorado, New and old Mexico, and all the Pacific States and Territories, met in one conglomerate mass. Virginia City , at that time the leading town, literally speaking, was but a den of highwaymen and murderers.

As a matter of course, a large majority of its citizens were respectable and law abiding, but nevertheless, there probably never was a city of the same number of inhabitants, whether situated in the uncivilized region of the Alps, or the sandy wastes of the Old World, or amid the hostile tribes that infest the auriferous region of the Andes of South America, or upon the scorching plains of Mexico, that represented so much crime as did this little city.

Here the feuds met. Those from the Pacific arraying themselves against those from the East; and thus the long winter nights were passed in debauchery, gambling and murder, resulting for a time, more to their own injury, than to outsiders, as scarcely a night passed, but some one of the ruffians were killed.

Little attention was paid to this class of murders, by the respectable portion of community, they being considered as a blessing to the country, rather than otherwise, so long as confined to their own ranks. Emboldened from the non-interference of the citizens, they soon organized themselves into a banditti, and commenced depredations on a broader scale, sending their emissaries to all desirable points in the mountains, to prey upon defenseless travelers.

So reckless, indeed, did they become, in their wholesale robbery, that not unfrequently, boys of the age of twelve or fifteen were robbed. One instance, in particular, is related, in which a boy of the age of twelve was robbed of seventy-five cents, by the notorious George Ives. This being all the money the boy had, was taken by this heartless wretch, who severely threatened the boy, stating that in case he should recognize him at any time, and expose, or attempt to expose him, that he (Ives) would kill him instantly, etc.

During the month of December, a Dutchman who had been mining at Virginia City, went down to Hillman's Ranch , which is situated upon Stink Water River , some twenty-five miles from the city, for the purpose of getting his mules that had been grazing there. Near this ranch, George Ives and several other desperadoes had a rendezvous, where they secreted their plunder. They had taken up a tract of land at the mouth of a deep canon, which formed an opening for several miles back among the huge and precipitous mountains of this range, and under the guise of ranch keeping, made it the base of their more desperate operaitons. As we shall have occasion to speak of this locality hereafter, we now return to Hillman's Ranch, where, upon the arrival of the unfortunate Dutchman, Ives and a fellow who went by the name of Long John, had casually called.

While settling with Hillman, the Dutchman carelessly exhibited some gold coin, of which he had about three hundred dollars. This was a tempting prize to men inured to crime, and as their piercing eyes fell upon it, as he was packing it away in his belt, they determined to rob him while on his return to the city.

Mounting their mustangs, they bore away to the southeast, across the broken prairie, passing behind a lone range of bluffs. Here they tacked to the southwest, striking the trail upon which the Dutchman was traveling, a little in his rear. Applying the spur to their steeds, they soon came up with their victim.

Ives being in the advance, drew his revolver, and

cried out to the Dutchman to halt. The Dutchman now seeing his danger, put spur to his mule, and dashed away at full speed, hotly pursued by Ives, who opened upon him a running fire, resulting in striking him in the back of the head with a ball, which brought him to the ground, a lifeless corpse. After having taken from his person what money and other valuables he had, consisting of a watch and revolver, they dragged him to a neighboring creek, submerging him beneath a parcel of old logs and float-wood, with a view to secrete their crime. They then made their escape, with the mules as part of the trophies of the morning, back to their rendezvous. The non-return of the Dutchman created suspicions as to his safety, in the minds of his friends in the city, and accordingly a company started out to learn his whereabouts. On arriving at Hillman's Ranch, they obtained information that soon led, not only to the discovery of the crime, but also to the apprehension of the murderers.

Up to this time, a consciousness of insecurity existed in the minds of all in this locality, and all felt the necessity of union in action, upon the part of the law and order portion of community. Crime had gone unpunished, although robberies had been committed, and men had been brutally murdered, in all sections of the mines, while little or no effort had been put forth to stay this high-handed crime, simply because resistance to their hellish work was rewarded with death. Months of pillage and outrage had, however, taught the people that security to property and life existed only in organization, and hence, secretly, but slowly and surely, an organization had been in progress, which was now counted by hundreds.

72

Secret detectives were appointed, and put upon the track of Plummer, and other leading characters. Committees of investigation were also appointed, yet with such caution did they move, that the existence of the organization was known to none save its members.

Ives and Long John were now brought to Nevada City , which is located two miles from Virginia City, for the purpose of being tried before the minor court of that city. The Vigilance Committee , as yet, not having obtained sufficient evidence concerning the gang, to warrant them in making a general demonstration, therefore concluded not to interfere with the trial. They, however, engaged Mr. Saunders to act as an attorney, upon the part of the people, while the murderer and his friends retained one Therman to defend him.

The prisoners, in the meantime, had been kept separate, Long John being considered the weaker vessel of the two, had been opportuned to make a confession, but upon finding all mild efforts futile, they took him secretly to a convenient place, and putting a rope around his neck, with one end over the limb of a tree, slightly elevated him, which was repeated the fifth time, when he was nearly strangled to death.

This, however, brought him to his senses, and acting upon the principle that "discretion is the better part of valor," made a full confession, disclosing the existence of the organized banditti, its form of organization, number of members, and its leading officers, among which, and the most conspicuous, was Plummer, who was its chosen captain.

During the trial of Ives, members of the gang hung

around the courthouse, apparently taking a deep interest in the proceedings. Anonymous letters were received, both by Attorney Saunders, and the court, setting forth that Ives had many friends, and if hung, the court, attorney, and all connected with his trial and execution, should suffer the consequences; that an organization of three hundred were sworn to protect him, and that all connected with his execution would not live twenty-four hours.

These letters, coming in just at this time, corroborated the statements being made by Long John, and the Vigilance Committee determined to push their investigations to the utmost extent possible. The letters were read by Saunders, to the assembled crowd, one of which was recognized as being in the handwriting of Plummer, read as follows, and was a fair sample of all received:-

Virginia City, Idaho.

Attorney Saunders—

Dear Sir:- This note is to inform you, and also the court, before whom Ives is now being tried, that he has many friends, who will not allow him to be hung, if possible to prevent it. He is innocent of the crime for which he is being tried, and if executed, an organization, consisting of three hundred and upwards,—who are sworn to protect him,—will avenge the outrage, and we warn all who are, and may be connected with the trial and execution, that they will suffer the consequences,

BY MANY FRIENDS OF IVES

After which the trial proceeded, resulting in the conviction of Ives, who was accordingly hung. Through the whole trial, he evinced but very little anxiety, apparently indifferent as to his situation, and when asked if he had anything to say why the sentence of death should not be passed upon him, replied, "that he had not, and for them to go ahead and do their damndest, he was ready." He evidently expected to have been rescued by the gang, which was probably the cause of his apparent indifference.

While in conversation with a friend, just before his execution, he said that, with "regard to his past career, he had nothing to regret; and that his only desire for the future, was to become as smart as Slaid was." When asked how smart that was, answered, "smart enough to carry about him a pocket full of men's ears." The trial having closed, the jury gave in their verdict, and the sentence of death was pronounced by the court.

A temporary gallows was immediately erected in a neighboring gulch, when Ives, accompanied by the sheriff, and others who were deputized for the occasion, moved forward, surrounded by a large assemblage of miners, who had convened for the purpose of putting down any attempt upon the part of the desperadoes to rescue Ives; but fortunately, no interference upon their part was made, and the execution passed off quietly.

Long John was still in custody, and now, after having been forced to a confession, talked freely with regard to the banditti; said that Sheriff Plummer was its chosen captain, and that upon his person papers might be found that would verify every thing he had said with regard to the gang. He disclosed the names of several leading members of the gang, who were then in Virginia City, at the same time giving the names of others in various parts of the country. From him much information was obtained, with regard to several robberies that had occurred during the preceding summer and autumn. A log store had been robbed in Virginia City, of twelve hundred dollars in gold and silver coin, besides several hundred in gold dust. The store had no floor except the ground,— which is usually the style in a mining town,—and the thieves had dug under the logs at one side, and reached under, had got hold of the box containing the money, which sat under the counter, and drawing it out under the logs, made their escape in safety.

Sheriff Plummer, who was at that time at Bannock City, was immediately notified, by the merchant whose money had been stolen, of the robbery, and that one Dutch John was suspected of having committed the crime, and that the said John was on his way to Salt Lake City, and would pass through Bannock on the stage, the following day.

On the arrival of the stage, Plummer arrested him, and found in his possession about four hundred dollars in coin and dust, which he took, then locking him up in his office until the next day, when he returned the money, and let him go.

As to this robbery, Long John's confession points directly to the complicity upon the part of Plummer, in the whole matter, though indirectly. He says, "Plummer caught the right person; that Ives was also interested in it, and that Plummer knew all the particulars connected with the robbery; that Dutch John informed him after his arrest, and reminded him that he was sworn to assist and protect him, whereupon Plummer released him, and reported to the

76

merchant that the Dutchman had no money that could be identified, and hence he had let him go."

We should say in this connection, lest the reader should confound the two, that this Dutch John is not the one spoken of in connection with the robbery of the stage between Bannock and Rattlesnake Ranch. This one lives in Salt Lake City, where he has two Mormon wives, and from whence he visits the various mines in the surrounding country, giving his time to gambling, and stealing, and to whatever else is necessary to accomplish his ends.

A plan was, at the time of Long John and Ives' arrest, being made for the robbery of Oliver's Express Office at Virginia City, and several merchants of the same place, and also George Chrisman's, and McDonald's stores, at Bannock City, which would, in all probability, have been successful, had the Vigilance Committee not been successful in obtaining the confession of Long John. Measures were immediately adopted for the purpose of apprehending the leading members of the banditti, not only in Virginia City, but in all the mining camps and towns of the surrounding country.

The Vigilance Committee had already tried several of the leading members of the gang, and sentenced them to be hung. Here we should say that the Committee always held their examinations in secret, and executed in public. Their form of trial was as follows: The Committee had a captain selected by ballot, whose business it was to preside at all meetings of the Committee. When complaints were entered against any one, which could only be done by members, outsiders, of course, knowing nothing of the organization up to this time, and investigation was immediately made by the Investigating Committee, which was appointed by the captain for that purpose , after which they reported in full to the Committee, when in session, the question then being taken up and discussed in full, preparatory to a vote which determined the disposition to be made of the parties upon trial, whether to be executed, banished or flogged .

When any one was condemned to suffer death, measures were immediately taken by the Committee to apprehend the party so condemned, and execute its decrees. A large number having already been sentenced to be hung, the Committee proceeded to its task with great caution and deliberation, and ere the desperadoes were aware of it, the city was surrounded and beleaguered by armed men, which rendered escape as impossible as it would have been had Grant's army formed a corden of bayonets around the doomed city.

"Alas! what stay is there in human state,
Or who can shun inevitable fate?
The doom was written, the decree was passed
Ere the foundation of the world were cast."

So effectually was every street and pass-way leading from the city guarded by armed men, that it was utterly impossible for any one to make their escape, while any that came were permitted to enter the lines, and pass into the city.

This concert of action upon the part of the miners, who had, one after another, been initiated into the secrets of the Committee, was the result of skillful management upon the part of its leaders, who had, at the same time, a large force detailed to search the

Execution of Five Desperadoes, at Virginia
City.

city for the doomed victims. Jack Gallager, Boon Helm, Hays Lyons, Charles Skinner , and a fellow by the name of Red, were arrested, all of whom had been previously condemned by the Committee.

These notorious roughs were now marched, one after the other, into P. Fout's store , where an officer of the Committee read the sentence of death to each, as they passed in. Here they were guarded until a temporary gallows could be erected, which was immediately done in front of Mr. Wilson's store, in this wise: A post was erected in the street opposite, and at a reasonable distance from Mr. Wilson's sign-post; a cross-beam was then made fast from the top of the sign-post to the other, and the victims brought forward.

Standing upon the scaffold erected for the occasion, with hemp nooses that were soon to environ their necks, dangling above their heads, and surrounded by an enraged multitude, yet unshaken and complacent, they stood, as though it was but an every day occurrence, although in the very jaws of death, and when asked if they had anything to say before their execution should take place, Boon Helm jestingly, and as a last request, asked for a glass of liquor, which was immediately presented to him, whereupon he turned to his comrades and drank their health, then returning the glass, he again turned upon his heel, and said to his comrades, "boys, I'll meet you in hell in five minutes."

Whether he kept his promise, or not, the writer is unable to say: but one thing we do know, and that is this, that during the following five minutes, their hands were tied behind them, the above mentioned nooses hastily adjusted about their necks, the scaffold knocked from under them, and the five, side by side, were swinging between the heavens and the earth, while their souls were launched into the shoreless regions of eternity, and thus ended the bloody career of five of the most reckless characters that ever infested any country.

Skinner had formerly kept a saloon at Bannock, but was at this time keeping a gambling and drinking hell of the darkest dye, at Virginia City, where men were murdered and robbed almost every night. He was an old Californian, where he had trained with Plummer, and been educated in the same school of vice.

Jack Gallager was formerly from Colorado, where he had darkened the annals of crime by committing some of the most atrocious murders known to that country. Boon Helm and Red were Western Idaho, while Hays Lyons was the said person, who in company with Buck Stinson, murdered the Deputy Sheriff of Virginia City, of which circumstances we have previously spoken.

Skinner, when arrested, was dragged from the bed of his wife, against her cries and entreaties; but this was no time to listen to the petitions of woman, and she was politely informed that in half an hour's time she could again see her husband. This happy twain had but recently been married. She had formerly lived with a man in Colorado, with whom she had emigrated to Idaho, during the summer of '63. They had settled upon a ranch in Horse Creek Valley, some fifteen miles from Bannock, upon their first arrival in the country, from whence she had visited Bannock, where she met Skinner, who soon charmed her with his wiles and gold, the latter, of which he freely gave, to robe her in fine clothes, and costly jewels. He was soon rewarded for his attention, by her in person, as she soon severed the brittle cord that bound her to her former master, and joined him as his miss.

Her former master followed her, and remonstrated with Skinner, but to no purpose, further than eliciting severe threats from Skinner, that, provided he did not desist his annoyances, he would get his d-d head blown off. This was rather cool treatment to receive at the hands of his rival, but as he was not "on the shoot," as the saying is, and knowing that Skinner was, he therefore concluded to let the wayward miss depart in peace, rather than attempt to coerce her, and hence made no further efforts to win her back from her late captor, who soon removed with her to Virginia City, where he subsequently, after having lived with her several months, made her his lawful wife.

Now to return to the subject: The half hour having expired, she was informed, by a wag who silently and unauthorizedly stole away to her cabin, that she could now see her husband. Stepping forth into the street, she beheld him hanging by the neck; when, with a frenzied and heart-sickening shriek, peculiar to her sex, she exclaimed, "For heaven's sake! have you killed my husband? What shall I do?" Thus continuing to give vent to her feelings in the most bitter wailings, until accosted by a fellow whom she knew, in something after this wise:

"Maggie, for God's sake, what are you making all this fuss about? Hush your noise, and stop your wailing over the loss of one man, in these times."

"Why, for the Lord's sake, how can you talk so, Dick? Don't you see my dear, innocent husband hanging there?" Again wringing her hands and bursting into tears of fitful agony, as if her very heart was breaking.

"Certainly," responded Dick, "and trust I as deeply sympathize with you as any one can, but this is no time or place for the exhibition of feminine weakness, in the way of cherishing love and regard for any one, though it be a companion, especially if it be one who has so forfeited his right to life, as to become a victim for the executioner,—and besides all that, 'there are just as good fish in the sea, as ever were caught.'"

Looking him in the face, she complacently exclaimed, "I believe you are more than half right; but it seems so hard to have him murdered in this brutal manner."

"Oh! nothing now can please me;
Darkness, and solitude, and sighs, and tears,
And all the inseparable train of griefs
Attend my steps forever."

"Of course I am right; and as to the 'brutal manner,' that we may talk over hereafter, but for the present, you should be looking after his effects, rather than mourning over his dead body, or first you will know, some one will have got his purse into their own hands, and you will be left out in the cold. In this country you know it becomes one to look after their own affairs, and as the Vigilance Committee has undertaken to dispose of him, they will unquestionably do it without any of your assistance, and my advice to you would be to look after matters that more immediately concern yourself."

"Oh, Dick, you would have me act upon the principle that 'the hide and tallow of an old ox will buy a young steer,' would you?" So saying, she dashed away to her cabin, and manifested no farther exhibition of grief. The next day, however, found her acting the part of an administratrix 'in her own right.' After settling up the affairs connected with her husband's estate, she realized a snug little fortune, which, as a sequel to the foregoing colloquy, she is now enjoying with Dick, in a beautiful little cottage, in the suburbs of Virginia City.

CHAPTER X

Many in the various towns and settlements of the country having been implicated by the testimony of Long John, an effort was therefore made by the Vigilance Committee, for a general demonstration to take place at several points, upon the same night; and hence, while the work of death was going on at Virginia City, the more peaceful town of Bannock was not exempt from its devastating and unwelcome presence. Three of the most noted characters of the gang were arrested, and summarily hung by the Vigilance Committee, among whom, and the most prominent, was Plummer, the captain of the banditti.

The circumstances of their arrest and execution, were as follows: The somber shades of evening had veiled the little city of Bannock with her dark mantle, and all save those who pass their nights in dens of infamy, and the Vigilance Committee, which, upon that evening held a special meeting, had retired to their respective places of abode, and all was quiet as

85

an autumnal eve in any rural New England town.

The Committee at this place was led by Captain S , (whose real name we decline giving, as he is still captain of the Committee). Arrangements for the arrest of Henry Plummer, the captain of the banditti, and Buck Stinson, who was one of the murderers of the Deputy Sheriff, at Virginia City, of whom we have before spoken, and also Ned Ray, a notorious bogus gold dust manufacturer, were made by the Committee; and as they emerged from their secret hall, they formed into three squads, and separated, the one led by Captain S-, crossing over the creek to Yankee Flats, for the purpose of arresting Plummer at his boarding house; while the other two companies, led by subordinate officers of the Committee, made their way, one to the saloon of Stinson, and the other to the cabin of Ray.

On reaching Plummer's boarding house, the Captain knocked at the door, when the lady of the house, who, by the way, was a sister-in-law of Plummer , opened the door, where, upon seeing her house surrounded by armed men, and a half dozen or more at the door, she uttered a shriek, and threw herself back upon a chair, as if frenzied at the sight. The Captain, at the opening of the door, had observed Plummer seated before the fire, in his shirt sleeves, (he having laid off his coat preparatory to going to bed,) and with a bound rushed into the house, followed by a number of his men. Plummer, in the meantime, made a rush for his revolver, which was lying under his pillow, as it was his custom to keep it by him nights as well as days, not knowing at what hour of either day or night his dreams might be disturbed, and his capture attempted. As he was about to seize his "faithful friend," several revolvers, and double-barreled shot guns were levelled at his head, with a shout from the Captain to stand, or he would be blown to atoms. Seeing himself surrounded with armed men, and with no possible chance of escape, he quietly surrendered. His sister-in-law, who, from the first sight of the men, had been much frightened, and had been screaming and crying, "that they were all going to be killed," now came up to Plummer, and throwing her arms around his neck, exclaimed, "Oh Henry, they are going to kill you!" to which he responded, "No, no, Ally; it can't be possible, for I have done nothing to merit death!"

He was then compelled to adjust his coat and boots, after which he was dragged from the agonizing embrace of his sister-in-law, and hurried off to the gallows.

Arriving at the bridge which crossed the mill-race, in the north portion of the town, they came up with the two other parties, that had been after Stinson and Ray, all converging at this point.

The meeting of the three squads of men at this place, each in charge of a rough, and all apparently moving towards a certain point,—namely, the old gallows, formerly erected by Plummer, for the execution of Horen,—seemed to indicate to Plummer the mission of those who had him and his comrades in charge, and for the first time, he fully appreciated the circumstances under which he was placed. Hitherto he had always been considered one of the bravest of the brave. In no critical position, where it required unyielding bravery and great daring, had he ever been placed and "found wanting," but now he saw his doom inevitably fixed. Just before him was the gallows,—beyond which was the dark and unexplored region of eternity.

The programme, with him was now changed. During his past career, he had found it an easy task to shoot down and butcher any who might cross his track, but always taking good care to get "the drop on them," as he termed it, and never allowing any one to get the first shot at him; thus making it a point to always fall upon his victim unawares, and killing him without any opportunity for defense. He had thus obtained a notoriety as being brave and bold; but now that the muzzle of the revolver was placed at his own head, ready to belch forth in tones of death, and carry the winged ball to his own heart; and the gallows, too, with all its horrors, its dangling noose and airy scaffold, with a just retribution beyond, his bravery weakens, and trembling with fear before his inextricable fate, he exclaimed, "My God! are you really going to hang me?" to which the captain responded, "We most certainly are." Plummer then burst into tears, and cried like a child; begging in the most pitiful manner for mercy, promising to leave the country, and never return, if they would but let him go. Finding that there was no hope of escaping his doom, he then begged time to write to his wife, which was refused. The Captain stating that "he had not given others time to write, and he could not expect more than he had given." He then asked time to pray, which was granted; kneeling down in the midst of the crowd, he sent up one of the most agonizing prayers that could

88

Execution of Plummer, Stinson and Bay, at Bannock.

be made, after which he said that he did not think God would ever forgive him, as he had committed so many crimes. Bowing himself before his captors, he again appealed for mercy, and for time to write to his wife. The Captain again repeated what he had already told him, "That he had not given others time, and he could not expect it; and furthermore, they had no time to waste with him;" so saying, he gave the command of "forward, march!" at which the crowd moved on, dragging their victims along.

Stinson and Ray had remained quiet, while Plummer had been making such a display of weakness; but the matter with them now began to assume a more serious nature, and as the gallows appeared, but a few rods in the distance, Ray began to hang back, finally throwing himself down, at the same time declaring that he would "be d-d if he was going to be hung." This plan did not work extra-ordinarily well for him, as the crowd did not drag him far, before one of the Committee stepped in front, and adjusted a rope around his neck, which was no sooner accomplished than a half dozen stalwart miners seized it, and with very little exertion upon their part, Ray was very materially assisted on his way to the gallows.

Arriving at the place of execution, the other end of the rope being thrown over the beam of the gallows, he was immediately raised to a proper height from the ground, and the rope made fast to one of the standards of the gallows; after which, Plummer and Stinson were served in like manner. Ropes having been made fast about their necks, were thrown over the beam, and the "hand over hand" process applied, until they reached an equal height with Ray, when all were left dangling upon the verge of that awful abyss, "from whence no traveler returneth," for their own benefit, but more especially for the benefit of the community at large.

Stinson and Ray appeared stubborn and willful until the last, while Plummer had sobbed like a child, with occasional shrieks, similar to those of a maniac, whose disordered imagination has coiled him with scorpions, and surrounded him with hideous devils, when the rope was placed around his neck. Stinson had, however, managed to get the knot under his chin, so that the rope did not choke him, and immediately after being suspended, commenced to sue for mercy, saying if they would only let him down, and save his life, he would confess all; while he was thus imploring for mercy, and begging for his life, an Irishman who was standing near the gallows, and who happened to be provided with a cane, gave him a thrust in the ribs, with the exclamation, "To hell wid ye's, we know enough o'ye's now." This specimen of Irish wit produced a burst of laughter from the crowd, notwithstanding the solemnity of the occasion.

Stinson's entreaties were unavailing, and he was compelled to close his hellish career, by a slow and lingering death of two hours upon the gallows, before life became extinct. Members of the Committee remained until all were dead, then retired, leaving the bodies of the three wretches suspended between the heavens and the earth until late the next day, when they were taken down, and placed side by side upon the floor of an unoccupied store on Main street, the doors being left open, so that all could see them who wished.

The following night, the Committee arrested a man who went by the name of Dutch John, (the one formerly arrested by Plummer, for robbing a store at Virginia City,) and escorting him to the store where the dead were reposing, placing a rope around his neck, and putting the other end over a beam, directly over the corpses, he was then suspended, and allowed to hang there several days, when all four were taken and buried in a lonely gulch,—called Murderer's Gulch,—half a mile from town.

"Sure, 'tis a serious thing to die, my soul!
What a strange moment must it be, when near
The journey's end, thou hast the gulf in view;
That awful gulf no mortal e'er re-passed,
To tell what's doing on the other side!
Nature turns back and shudders at the sight,
And every life-string bleeds at the thought of
    parting."

It is proper in this connection to state, that upon Plummer's person were found papers confirming all that Long John had said, of both him and the banditti. A document similar to a Constitution and By-Laws was found, which had been signed by over eighty of the members of the gang.

It appeared from this document that the banditti was,—in a degree,—organized similar to a military company, with a captain, two lieutenants, and other subordinate officers. The captain and lieutenants were elected, while the minor officers were appointed by the captain.

Attached to this document, and just above the

92

signatures was an oath, which had to be taken by each applicant, before he was admitted into the secrets, and councils of the banditti. This horrid oath prescribed that each member of the gang should obey the orders of the captain, and other officers placed over them, whatever the work to be accomplished might be, or the consequences devolving upon its performance.

It also prescribed, that if any member should divulge any of the secrets of the band, that he should suffer death by first having his bowels taken out, after which his throat cut from ear to ear, and making it the solemn duty of any and all members to see that this was accomplished at the earliest moment possible.

After the reading of the oath, by an officer, to the applicant, he was compelled to fall upon his knees, in the midst of the gang, and with uplifted hands, call God to witness his solemn oath, after which, and while yet upon his knees, sign his name to the document, with a pen dipped into what was termed by the gang, "devil's ink," which was really blood taken from the veins of some murdered man, and kept in a bottle for that purpose.

Having passed the ordeal of initiation, the applicant was then instructed in the secret signs and grips, etc., of the banditti. Besides the grips and pass-words, they had a visible sign, or mark, by which a member could recognize another upon meeting him. This mark was a peculiar knot in the neck-tie, or cravat worn about the neck. It was called by the gang, the "cordon knot," and signified "a band of brothers, linked in eternal friendship."

93

This band of cut-throats had several places of meeting, one of which was at Skinner's Saloon, of Virginia City, another at Stinson's Saloon, at Bannock City, and a third at Picket's Ranch, on Rattlesnake. At these places they convened as the occasion required, to lay their plans for robbing and murdering, as the case might happen to be.

As to the other officers of the gang, the statement made by Long John, namely, that Ives was first lieutenant, and Jack Gallager, who was hung at Virginia City, second lieutenant, was corroborated by the document found with Plummer.

The information obtained from this document proved of great service to the Committee; not only was it valuable for its detail, but it had also furnished the names of its members throughout the whole length and breadth of the land. How promptly the Committee acted, upon receiving this information, and with what crushing effect upon the banditti, subsequent pages will show.

CHAPTER XI

The news of the wholesale execution of desperadoes
at Bannock and Virginia Cities, fell among the gang
like a bomb-shell in an enemy's camp,—with telling
effect,—and hence a general stampede from the
country, of this class, immediately took place.

"Any other mines, or sections of country being
preferable to this," as they said; some claiming that
"the mines in Western Idaho were richer than those
of this country;" others that "Arizona was the great
Eldorado." "Any other place at all but Eastern
Idaho." "The mines here have proved a failure, and
they were going to leave," etc.

All sorts of excuses, similar to the above, were
made by this class of men, as they hastily betook
themselves to other sections of country.

They had observed the unanimity in action, of the
miners and the law and order portion of community,
in the execution of those already arrested, and knew
there was an irresistible power behind, that would

95

sooner or later overtake them, and therefore sought safety in flight.

Many who had not the means for a long journey, secluded themselves in the neighboring mountains. Several of these poor wretches, after having suffered weeks of torture, from cold and hunger, came straggling, one after another, into the various settlements, with frozen limbs, and distorted and emaciated countenances, resembling more the relics of some ancient clan of devils, than of human beings. Their sufferings, however, were no palliation for their crimes, and they were "gobbled up" by the Committee, and summarily dealt with, according to their just deserts.

In the meantime, the Vigilance Committee had organized several small companies, numbering from ten to twenty-five persons, for the purpose of scouring the country, and picking up fugitives,—members of the gang,—who were making their way out of the country. One party, consisting of twenty-five men, was organized, and led by one Clark, for the purpose of pursuing several desperadoes, whom it was known had "skedaddled" over the mountains, for Western Idaho.

To cross the Snowy Range in mid winter was a very hazardous undertaking, and to any save those inured to hardships, and acquainted with the mountain country, it would have been considered impossible; but the men selected for the occasion were stalwart and brave men, who had, from years of experience in frontier life, become familiar with such expeditions, and the tragedies usually enacted on similar occasions, and hence they embarked in the enterprise with a will that carried all before them. Well armed and equipped, and mounted upon their steeds, with a supply of provisions to last them three weeks, they set out on their journey.

The afternoon of the third day found them encountering a severe snow storm on the main chain of the mountains, between the head waters of the Missouri and Columbia rivers. For miles, at the base of the mountains, they had been able to follow the trail of the desperadoes, who were flying for dear life, but as they commenced ascending the rugged steep, a snow storm set in, with a prior fall of two feet, which was now driven before the mountain tempest in every direction, not only blinding the trail, but rendering further pursuit almost impossible.

Steep acclivities, with huge rocks projecting, were encountered and overcome by the party. Thus they struggled on during the remainder of the day; night finding them buried in snow, exhausted with fatigue, and incapable of further exertion, they encamped in a ravine, where they found a supply of fuel. Huddling their horses together in a pine thicket, they passed the night without losing any from freezing, although the weather was exceedingly cold.

The following morning, led by their guide, who was familiar with the passes of this range, they soon pressed their way through the snow, and over the summit, from whence they soon descended into the beautiful valley of Deer Lodge river. Here they found country destitute of snow, with an abundance of grazing for their stock. After recuperating their animals, they swept down the valley, soon reaching Cottonwood City, where they came up with five of

the gang, surprising and capturing three, while two
made their escape to the mountains. Among the captured was William Bu ton, of whom we have before
spoken .... th b ... of the stage
.... The
parti ... ... on,
however ... ...ers
and robberies among which was that of robbing the
stage, as before mentioned, also of stealing and
butchering a Mormon's oxen, of which we shall speak
hereafter.

The other two, who had escaped, were hotly pursued by some eight or ten of the company, the chase
continuing some three miles, when, upon being surrounded, they threw down their weapons and begged
for mercy, but mercy was foreign to the mission of
their captors, and they were accordingly suspended
to the branch of a neighboring tree, and left to be
taken down and carried away by the buzzards and
ravens that infest the wilds of that country.

Rejoining the balance of the party, they broke
camp for Fort Owen, which is situated upon the west
side of the Bitter Root Mountains, about two hundred
miles distant. The Captain having learned, during his
stay in Cottonwood City, that a party of from ten to
fifteen of the band had passed through town only the
day previous to his arrival, determined to overtake
them before crossing the Bitter Root Mountains, and
hence, pushed down the valley with renewed vigor.

For two days he found the country well supplied
with grazing for his stock, with no snow, while the
mountains, on either side of the valley were clad in
their wintry livery. At three o'clock, P.M., of the third

98

day, they came up with the desperadoes, who were encamped at the juncture of the Deer Lodge, Fort Benton, and Fort Owen roads. So sudden and unexpected was the approach of Mr. Clark's party, that they decamped in confusion, a part taking down the Bitter Root Valley, while the balance dashed away over the Fort Owen road, hastily pursued by Clark and his party.

Reaching the Bitter Root Mountains, just as the shades of evening were coming on, they were compelled to abandon the chase for the night, and camp, lest they should be overtaken by a storm in the mountains. Early the following morning, resuming their line of march, the pursuit was continued over hills and through gorges; finally ascending the highest mountain of the range, they came upon the deserted camp where the roughs had stopped during the night. Here was witnessed a most heartrending scene. The flight of these members of the band, the day previous, had been so precipitous that they had not only left behind much of their provisions, but also their blankets, and therefore were compelled to pass the wintry night, which was intensely cold, without covering, and to render their condition more dangerous, and death certain, the mountains at this point were barren of timber, and thus, without food or fuel, with the drifting snow falling about them, they freezingly passed the night. Here, upon the beaten snow, lay two in the cold embrace of death, while but a short distance from them were three more frozen objects of humanity, in whom life had not yet become extinct.

Taking a hasty glance at the camp, the Captain observed that the surviving portion of the gang had continued their march, and hence ordered a forward movement, leaving the poor, unfortunate creatures to become living monuments of eternal ice and snow.

Soon descending the mountains, they reached Fort Owen, where they overtook the last survivors of this portion of the gang. Here they had put up, designing to remain a day or two, not thinking it possible for their pursuers to cross the mountains, where they had so narrowly escaped freezing.

Their capture having been accomplished, which took place in the Fort, while they were enjoying a supper, they were compelled to mount their horses. There being but three in all, and one of whom had so jaded his horse that the poor animal could scarcely move, was compelled to mount, and sit behind one of his captors. In this condition they were hastily marched off to a convenient grove, when all three were properly stationed under a projecting limb of a huge tree. Ropes were then adjusted about their necks, and made fast to the limb. All was now ready, and at the signal of the Captain, the two horses rode by the desperadoes, were led from under; at the same instant, the man who sat upon his horse with the third victim, put spurs to his steed, and galloped from beneath, and thus the three wretches were hurled into eternity.

The following morning, after having enjoyed a night's repose in the Fort, as well as the hospitalities of Captain Owen, who is in charge of it, Mr. Clark and his party set out for home, taking a different and more feasible route, by which they would enter the Bitter Root Valley some fifty miles below the juncture of the roads, where they had first surprised the latter

100

Execution of Three Desperadoes, Near Fort Owen.

gang of desperadoes. By taking this route, two points were gained, namely: First, they avoided re-crossing the Bitter Root Mountains; secondly, it enabled the party to intercept that portion of the gang that had sought safety among the inhabitants of the valley. On the second day, Clark and his party found themselves winding their way up the valley of the meandering Bitter Root, occasionally passing the lone inhabitation of the husbandman, before whose strong arm the forests of this country must give way, and the valleys bring forth.

Reaching a more thickly settled section of the country, they soon learned of the presence of two suspicious looking fellows, who had but recently made their appearance in the settlement, and who were,— as they were informed,—stopping at the cabin of one Williams This Williams, as Mr. Clark was informed, had located in this settlement in the early part of the preceding summer, and had been considered a suspicious character, from the fact that whenever any of like character were passing through the valley, they always stopped with him.

Acting upon this information, Clark and his party immediately surrounded his cabin, capturing both him and the other two fellows.

While marching them off to a tree for execution, a chap who went by the name of Yank, having Williams in charge, quietly suggested to him,—in a low tone of voice, so as not to be heard by the rest,—that he had better make a full confession; that he (Yank) knew the captain of the party, and knew that that was the only thing that could possibly save him from being hung.

As the ropes were being placed around their necks, Williams begging for his life, proposed to confess. Here the execution was staid for a few minutes, while he proceeded to state that "he had joined the banditti, when at Bannock, where he had passed by the name of Texas, the previous spring, and that, in company with Plummer, had murdered a merchant by the name of Granger, in Big Hole Pass, after which, and in accordance with the suggestions of Plummer, and others of the gang, had located in Bitter Root Valley, for the purpose of establishing a rendezvous there for the banditti."

He further stated that "the two arrested with him were members of the gang, but that there were no others in the settlement." His confession, however, did not save him, and all three were left dangling, as the party dashed away, homeward bound.

On arriving in Big Hole Pass, the bones of Granger were found beneath a mass of rocks in a deep gorge, as designated by Williams.

Soon reaching Virginia City, the Captain reported to the Vigilance Committee the execution, and destruction by frost, of sixteen members of the banditti, and thus ended this horrible and bloody tragedy.

CHAPTER XII

Few men have lived, who have left behind them as dark a record as has the notorious Slaid, whose atrocious deeds of blood mark the continent from the Missouri river to the Pacific.

In entering upon the biography of this wonderfully peculiar man, we do it with a degree of hesitancy, knowing that we must necessarily confine ourselves to a very small space, while the startling details, and shocking incidents of his life furnish sufficient data for a luminous volume.

Other men have, perhaps, startled communities with an isolated murder, equally as atrocious as any committed by him; but few, if any, have ever been permitted to continue so long a career of blood, uninterruptedly, as has he. Commencing his deeds of murder in Kansas, in its early days, and extending them across the wilds of the continent, always keeping beyond the pale of civilization; thus for years his name has been a terror throughout the Western country

105

Not like many desperadoes and highwaymen, whose thirst for blood originates in their insatiate desire for gain, but on the contrary, he would scorn to murder for money, yet for the slightest offense, though it came from his best friend, he would fall upon him, cutting him to pieces in the most fiendish manner possible, apparently more to satisfy an ungovernable passion of destructiveness than for any other purpose.

One little incident, which illustrates his whole character, and we pass on to his career at Virginia City. While in the employ of the Overland Stage Company, some years since, stationed between Denver and Salt Lake cities, in charge of a ranch, or rather stage station, it so happened that on a certain occasion his herdsmen could not find all the stage mules, which, by the way, were grazing but a short distance from the station, and hence the stage of that morning had to lay over a few hours. Slaid, in the meantime, accompanied by three or four drivers, set out in search of the missing mules, not traveling far, however, before they met three half-breeds (French and Indian) riding the animals towards the station.

These unfortunate men were herding a large stock of cattle on a neighboring stream, where the stage mules usually run, and on this eventful morning had taken the animals without Slaid's permission, to look up some strayed stock.

On riding up to the men, Slaid commenced firing away at them, with his revolver, the rest of his party standing at a little distance, apparently looking on. At this juncture, the three men, who were defenseless, they not being armed, dashed away toward the station.

They had not gone far before it was apparent that

106

one had been struck with a ball, as directly he reelingly fell from the animal. On riding up to him, Slaid exclaimed, "Steal another mule, will you, you d-d thief?" at the same instant dashing away, leaving the poor man to die, they soon reached the station, where he and his party captured the other two, immediately hanging them upon a temporary gallows, constructed for the occasion, in this wise: Two wagon tongues, or rather the poles from two coaches were taken, and chained together for the purpose of making a crotch, similar to what a farmer would construct upon which to suspend his swine, the day of butchering: it was then placed near the corner of the station-house, with a pole thrown across, and the two victims brought forward, and suspended by the neck.

Here they were allowed to remain, while Slaid and his party entered the house, and partook freely of a breakfast prepared by Mrs. Slaid, after which the bodies were taken down and buried, by the drivers stationed at this place.

We might enumerate incident after incident, where he has exhibited equal coolness in the prosecution of the destruction of human life, but we hastily pass on.

In the summer of '63, we find him located upon a ranch in a beautiful valley fifteen miles north of Virginia City. During the following autumn and winter, he visited Virginia City frequently, indeed, it was not an uncommon thing for him to remain in town for days, and even weeks, during which time he usually kept intoxicated, rendering himself not only odious, but also dangerous to all with whom he came in contact.

Lounging about the gambling and drinking saloons,

107

with a huge bowie knife, and two revolvers strapped about his waist,—with a reputation dark with crime, that had gone before him, it was not strange that men of soberer taste should shrink from his presence as he passed up and down the streets, entering saloons, stores, and private houses, brandishing revolvers, insulting and threatening whomsoever he saw fit. In fact, he seemed to take a sort of hellish delight in galloping down the street on horseback, suddenly wheeling up in front of a store or saloon, as the case might be, and with a thrust of his spurs into his poor animal's ribs, dash through the door, parade his horse about the store, or in front of the bar—if in a saloon—hallowing in the most boisterous manner, and swinging his revolvers, for the purpose of frightening all who might happen to be in the room, which was generally the result, all "skedaddling," either through the back door, or windows, or skulking behind the counter, after which he would force his animal either out at the front or back door, and dash away to some other place, and go through a like performance.

Thus for months he kept the people awe-stricken, no one wishing to run the risk of his life in opposing him, knowing that certain death would follow, unless they should get the "drop on him," and first give him a mortal stroke. But finally, after months of toleration upon the part of the people, a complaint was entered before the miner's court of the city, and papers issued for his arrest. The sheriff at once repaired to the saloon where he was, and stepping up to him, notified him of the issuing of the papers, and that he was his prisoner, whereupon Slaid requested the privilege of seeing the papers, which was granted by the sheriff,

Slaid instantly tearing the papers into atoms, at the same time damning the court in the most bitter language, and with a bold push forced his way out through the door, mounted his steed, and galloped away to the court house, entered the court room, and with curses, threats, and waving revolvers, broke up and drove the court out of the house.

This bold attempt upon the part of a desperado to break down civil law, was more than the Vigilance Committee could stand by and quietly submit to, and hence measures were immediately adopted by its leading members for the purpose of bringing about his arrest and execution, and scarcely an hour had passed subsequently to the breaking up of the court, when the city was beleaguered by armed men, and his arrest accomplished. He was at once disarmed, and marched off, surrounded by armed men, to a convenient place for execution.

On arriving at the gallows he begged for mercy in the humblest terms, but on being assured that they were determined to hang him, he called for Judge Davis, whom but a few moments before, he had driven out of the court house, and earnestly implored him to intercede in his behalf; but when told by the Judge that he could do nothing for him, he then called for Attorney Saunders, who came forward, and also assured him that he could not save his life, as the people, and not courts were adjudicating his case.

The victim then plead in the most suppliant and earnest manner for them to spare his life until he could see his wife, stating that "he had business matters to settle up, which she could not do without great loss to herself, if he was thus taken from her, without first having given her the required information." His wife at this time was at home, fifteen miles from town, but a messenger had been dispatched by one of Slaid's friends, immediately upon his arrest, and notwithstanding the Committee were aware of that fact, yet they declined to await her arrival, knowing that Slaid's real object in having her present, was simply to create sympathy in his favor, in the minds of the crowd, and thus procure his release. Therefore he was denied the poor boon of seeing her, and when positively informed to that effect, he exclaimed, "My God, my God! must I be murdered without seeing my dear and beloved wife?"

His hands were now tied behind his back, and the noose placed around his neck, and as its sleek coils touched his skin, sending a chilling, death-like sensation throughout his whole system, he cringed and writhed as though pierced with the deadly fangs of a legion of scorpions, and shrieked in the most bitter agony for his dear wife. "Let me see my wife, my beloved wife." At this juncture the crowd, which by this time had become very large, gave signs of a disposition to interfere in his behalf, and cries of "let him see his wife," came from all quarters of the assemblage. An instant more, and four hundred guns were leveled at the crowd, when all again became as hush as the stillness of death, and the execution proceeded.

Life having become extinct, his corpse was taken down and placed upon a bench in the Virginia City Hotel, to await the arrival of Mrs. Slaid.

Scarcely a half hour had elapsed after his execution, before she came galloping down the canon leading into the city; rounding the corner, and taking down

Mrs. Slaid Beholding the Corpse of Her
Husband.

Main street, she soon reached the hotel, having learned that her husband was there, she dismounted, and hastily stepping into the house, not yet knowing that he was dead. On beholding his lifeless corpse, still apparently fresh with the tints of life and health, she threw up her hands and burst into tears, then falling upon his pulseless form, clasped her arms about his neck, and in a long and lingering embrace held his once proud and stalwart form, weeping and sobbing, with occasional bursts of lamentation, as "My dear husband, you are not dead, it can't be possible." Again, "They have killed you, they have killed you, my beloved husband," etc.

> "My grief lies all within,
> And these external manners of laments
> Are merely shadows to the unseen grief,
> That swells with silence to the tortur'd soul."

Finally she arose, and with an air of revenge, exclaimed to one of the bystanders,—who, by the way, was one of Slaid's intimate friends, and in fact was in his employ, it being his usual custom to have one or two of his class of men accompanying him whenever he was dissipating, or likely to get into a row,—"Why did you not shoot him, and not let them hang him like a dog?"

The ruffian whom Mrs. Slaid thus addressed, had, but a few minutes previous to her arrival, made threats with regard to his executioners, when he was seized and hurried off to the gallows, and but for the timely interference of one or two leading men, would have shared the fate of Slaid.

It is due to Mrs. Slaid, to say in this narrative, that while it is said of her that "she was as desperate as

112

her husband, and will shoot as readily as any man, and that she has already killed several persons," yet she is a woman of more than ordinary abilities, and by the casual observer, would be considered a lady of refinement.

It was the fortune of the writer, during the fall and winter of '63 and '64, to dine several times at their table, and we must say that the impressions made upon our mind, with regard to them, would have been very favorable indeed, had we known nothing of the parties; and notwithstanding they were such desperate characters, they regarded each other with as much esteem and tender affection as any twain could, though the most refined.

One word with regard to Slaid, and we leave him to repose forever, among the broken hills and desert wastes, where he has so often besmeared his hands with human gore, and startled the councils of pandemonium with his atrocious deeds. It could not be expected that a man of his character would pass through such a career, unscarred with the bowie knife and the bullet, nor was it the case with him, for his person bore many external marks of the knife, indicating many hand-to-hand fights, while at the time of his execution, and for some years previous, he carried, buried in his flesh twenty-one bullets and buckshot, of which fact he often boasted, declaring that it was impossible to kill him with lead.

CHAPTER XIII

MURDER OF MCGRUDER, CANDIDATE FOR CONGRESS, AND
GEORGE COPLEY, OF BANNOCK CITY. EXECUTION
OF SEVERAL DESPERADOES.

While the work of death was progressing in Bitter
Root and Deer Lodge valleys, it ceased not to "stalk
at noon day," as well as at midnight, at Bannock and
Virginia cities, and many a poor, misguided victim
found the termination of all things earthly, in the
hangman's noose.

A few days only, subsequent to the execution of
Captain Plummer, a Spaniard, who had made himself notorious as a highwayman and murderer, was
condemned and sentenced by the Committee at Bannock, to suffer death upon the gallows, and accordingly measures were adopted to enforce the decree,
a part of which was the dispatching of ten men by
the committee for the purpose of arresting the doomed
victim.

This small party was led by Mr. George Copley,
a highly respectable citizen of Bannock City, who
after having stationed a guard about the cabin, where

115

the Spaniard was stopping, entered the door, accompanied by two or three others.

The Spaniard had evidently been on the look out, expecting, perhaps, that it might be his turn to swing soon, and hence, upon the first appearance of the party, secreted himself under a bunk in one corner of the room, and when Copley and his comrades entered, commenced firing upon them, shooting Mr. Copley through the abdomen, which resulted in his death, some ten hours subsequently, also wounding another man in the hip, but not fatally.

This was the first unfortunate event that had occurred with the Committee in its work of death and general purification, and hence great indignation was felt by all good citizens. Immediately upon his being shot, a large throng of miners and townsmen assembled at, or near the cabin. Mr. Copley was carried away, and a small howitzer, which Captain Fisk had left in the city the previous autumn, was at once brought forward, and a few shell thrown into the cabin, the explosion of which not only slightly elevated the roof of the cabin, but also prostrated the Spaniard, who was immediately suspended by the neck, to a pole thrown across the corner of the cabin.

Scarcely had his feet left the ground, when fifty or more revolvers were drawn, as if by one common impulse, which sent their contents crashing through his already lifeless corpse, completely riddling his body into shreds. After which he was cut down, the cabin torn to pieces, thrown into a heap, and fired, consuming his mangled remains, which were thrown across the logs, and given to the flames. Thus ended one of the most dangerous men that ever infested our frontier.

116

Shelling a Desperado out of His Cabin at
Bannock City.

The work of the Committee progressed finely in all the settlements around about Bannock and Virginia cities. Several, in various sections of the country had made confessions, all agreeing as to the existence of the banditti, which fact, taken in connection with the information derived from the papers found with Plummer, removed all conscientious scruples from the minds of the Committee, and all whose names were attached to the document, were provided with a hemp cravat, with a different knot (hangman's knot) from that worn by the banditti. Among those executed before the writer left the territory for the States, were Henry Plummer, captain of the banditti, George Ives first lieutenant, and Jack Gallager, second lieutenant, Buck Stinson, Edward Ray, Frank Parish, Frank, (a Spaniard), two men who went by the name of Dutch John, also a man who was called Red, and a man by the name of Brown, Boon Helm, William Bunton, Stephen Morrison, Charles Skinner, Aleck Carter, Bob Zachery, Bradley, W. Graves, George Thermer, Bill Hunter, John Cooper, Slaid, Rowley, and various others, whose real names were not known.

Those banished were L. Therman, the attorney who defended Ives, George Kuster, H. P. Sessions, George Hillman, H. P. Smith, who defended Stinson and Lyons at the gallows, also one Moyer, besides several whose names we could not learn. One man was also striped with fifty lashes, at Nevada City. The offense, though grave in its nature, was simply handing his revolver to Bradley, knowing that he wanted it for the purpose of killing a man, which he immediately did.

Three of the above named wretches were hung near

Ives' Ranch, having been captured at his former rendezvous, which is a secluded cave among the broken mountains, some three miles up the canon from his ranch, before spoken of. In this cave much stolen property, such as saddles, bridles, articles of merchandize, watches, etc., were found, and it is believed that much gold is buried there, but no traces of it have, as yet, been found.

One unprecedented murder, we had well nigh forgotten to mention, namely, the murder of McGruder, candidate for Congress. This unfortunate man was stumping the territory, or, in other words, he was traveling from one settlement to another, electioneering for himself, and while on his way from Bannock City to Lewiston, was waylaid, and shot from his horse, as were also two others, who were accompanying him, by three dastardly wretches, who had followed them into the mountains. His murderers were, however, pursued, and two of them arrested in California, and subsequently hung.

In our narrative we had intended to have spoken more fully with regard to theft, or larceny, but the subject of murder has been prominent in all the category of crime of that country, that we have well nigh ignored all minor matters, but there is one circumstance, so inhuman in itself, that we cannot pass it by. A Mormon boy, of the age of fourteen, had been intrusted with his father's freight team, consisting of two yoke of cattle, for the purpose of taking them from Virginia to Salt Lake City. When on the second day out from the former place, he was stopped by three desperadoes, among whom was William Bunton, his cattle unyoked, and three of the fattest taken and butchered on a neighboring bottom, the boy being severely threatened. This instance of outrage and pillage was but one among the many which occurred all over the country.

But well may the people now rejoice that this band of cutthroats is broken up, and that the millennium of devils has passed away forever, from the fair valleys, and mountain retreats of Idaho and Montana. before the light of civilization, that is laying the first frame work of civilized society across the continent. whereon will be built thriving and prosperous states. ere the present generation shall pass away. Thus has civil law been established in what is now Montana Territory, and which is now assuming the established forms of government. May God shield it as a Territory, and guide it as a State.

"Ah! why will men forget that they are brethren? why
    delight
In human sacrifice? Why burst the ties
Of nature that should knit their souls together
In one soft bond of amity and love?
Yet still they breathe destruction; still go on
Inhumanly ingenious to find out
New pains for life, new terrors for the grave;
Artificers of death!"

THE END